Eye of the Blackbird

A Story of Gold in the American West

Holly Skinner

Johnson Books
BOULDER

Published by Johnson Books, a division of Johnson Publishing Company,
1880 South 57th Court, Boulder, Colorado 80301.
E-mail: books@jpcolorado.com

9 8 7 6 5 4 3 2 1

Cover design by Debra B. Topping
Front cover photo of the author provided by the author.

Library of Congress Cataloging-in-Publication Data
Skinner, H. L. (Holly L.), 1960–
Eye of the blackbird: a story of gold in the American West / Holly Skinner.
p. cm.
Includes bibliographical references.
ISBN 1-55566-312-5 (alk. paper)
1. West (U.S.)—Gold discoveries. 2. Gold mines and mining—West (U.S.)—History. 3. Gold Miners—West (U.S.)—History. 4. West (U.S.)—History—19th century. 5. Frontier and pioneer life—West (U.S.) 6. West (U.S.)—Description and travel. 7. Skinner, H. L. (Holly L.), 1960– 8. Women miners—Wyoming—Lander Region—Biography. 9. Lander Region (Wyo.)—Biography. 10. Prospecting—Wyoming—Lander Region. I. Title.

F591.S594 2001
978′.02—dc21 00-054960

Printed in the United States by
Johnson Printing
1880 South 57th Court
Boulder, Colorado 80301

♻ Printed on recycled paper with soy ink

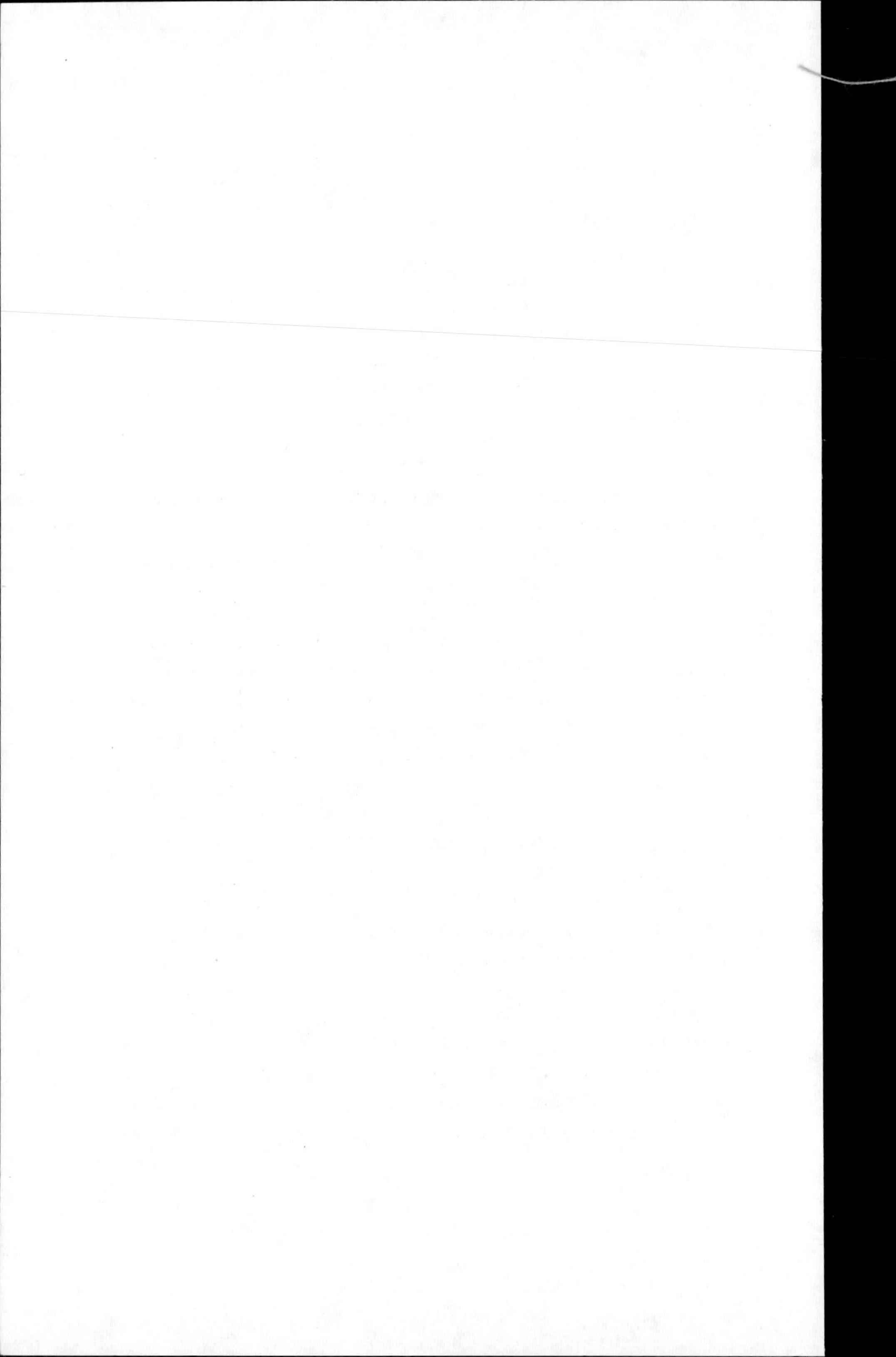

Introduction

When the blackbird flew out of sight
It marked the edge
Of one of many circles.
from Wallace Stevens
"Thirteen Ways of Looking at a Blackbird"

I'VE BEEN CROUCHED by the water for an hour, spinning sand like an incantation. Four empty sample bags are crumpled on the stream bank, and I stand and stretch my tired muscles before I empty the fifth into my pan. The hat shading my eyes used to be a fine Stetson. The brim droops now and the crown, worn through at the creases, is carefully sewn back together. My clothing is also bedraggled and dusty, jeans tucked into rubber boots, and canvas shirt torn from too many encounters with barbwire fences. Wisps of sun-bleached hair have escaped from their braid to brush against my burned cheeks, and I try to wipe them back with my shoulder because my hands are still wet. I find I can't begin to care how I look—field-worn, mud-spattered, sun-baked—because I haven't come face to face with another human for weeks. If I set out to step backwards and away from this century, I have nearly fallen over.

I stand watching the cold river run by me, while long shadows of a dying October stretch toward the canyon's rim. The cliffs rising suddenly from water's edge have caught the last light and glow in tones of gold. I bend and pour the fifth bag of dirt onto a screen over my pan. Each bag, filled and labeled and carried here to wash, represents a different dry gulch and a world of possibility. Crouching again, I dunk the pan and scrub the gravel with my fingertips until all the finer material has slipped through the screen. I study the larger gravel left behind, looking for likely host rock, evidence of deposition, and the occasional pleasant surprise. Finding nothing worth keeping I dump the screen with a splash into the river, and knead the dirt in the pan to raise the silt and break up lumps. I spin and swirl the contents, dumping a little at a time over the lip of the pan and scooping in more water in a sparkling gyration that suspends both sand and time.

A plume of silt spills out of the pan and into the bright water to be caught and carried by the current until its trail disappears downstream. It strikes me that this is what the nineteenth century was like when miners rushed from stream to western stream—a plume of silt. In once-clear water, suddenly there was digging and turbulence and noise. Then the noise began to fade and the diggers to drop out until the river of the world swept on to other things. They had been looking for gold, as I am looking for gold.

Sand washes over the edge of the pan until only heavy, black magnetite remains. I pour out all but a few tablespoons of water, then rock the black sand around the bottom so it will show its golden tail. This is the breathless moment, when expectation holds you spellbound. Often there is nothing to see but wasted effort. But this time I am lucky. In the tail of black sand there are tiny specks as brilliant as the sunlit cliffs, unwavering, luminescent gold.

I pick the flakes out with the tip of my finger and shake them into a small, stoppered bottle which slides back into my pocket with an unnatural weight. I must add that particular dry gulch to a long list of places where I've found gold, and I must add this day to the longer quest of humankind toward the same end.

How could a simple element like gold cause mountains to be moved, civilizations to rise in splendor and get torn back down, new worlds to be explored and populations to shift like cargo in an unsteady boat? Hundreds of thousands have joined in the search for gold, sifting earth and breaking stone, pillaging the living and robbing the dead, all to hold in their hands for a moment some small piece of glowing treasure.

In the American West, gold made "Manifest Destiny" not a theory, but reality. As the primary catalyst for settlement, gold was at the center of conflict over western lands. Red Cloud's War was a war about gold, and Custer's death was a death due to gold. Gold made young America rich and precocious. Anything was possible if you followed the rainbow's end. Anything was possible if you didn't go mad in the pursuit.

The psychology of gold has much to do with its innate qualities, its beauty and weight, purity and eternal nature. Gold has a value in and of itself, which made it useful as a medium of exchange for centuries. But gold demands balance and discourages growth for growth's sake, so it is no longer appropriate as money. For decades it was illegal to own gold in this country, such a subversive metal.

It has become easy in the light of today's "goldless" society to dismiss past centuries' gold rushes as a rapacious trampling of resources—greed run ecologically amok, an exploitation to be condemned and shunted into the closet with the rest of mankind's unpleasant deeds. But that view also dismisses the reality of human courage, the labor and hardship, the dreams and disappointments of those individuals who lived the history.

I have come here to relive that history, to walk at least partially in their footsteps, carrying pick and pan. For a year and a half now I have lived in a ghost town where miners once dug their hearts out, gave up and moved on. By kerosene lamplight, in a run-down log cabin on Wyoming's South Pass, I have read the diaries and letters, reminiscences and requiems of those who joined the western gold rushes. From California to the Klondike, Virginia City to Atlantic City, Denver to Deadwood, Clear Creek to Cripple Creek I have followed them—those dreamers and die-hards—mad alchemists wringing gold from sand and solid rock.

I walk circles around them in my mind, to see what their lives meant, to weave them into the larger picture of history. Their stories are like pebbles in a pond rippling outward, like a river running past cloudy then clear.

Reluctant to leave, I dangle my booted feet in the rushing water while the sun prematurely fades and a milk-drop moon rises in the pale sky. But I am telling you the end of the story when I should, as they say, begin at the beginning, thirteen months ago.

Chapter 1

THE WIND IS BLOWING HARD from the west, a dry, hot wind that's blown all summer. It is September in a year of drought and forest fires that black the air and redden horizons. You can see the wind on a day like this, twisting across a dry alkaline lake bed to funnel white powder two hundred feet in the air with the regular eruptions of a geyser.

I grind my pickup over another ridge of Wyoming sagebrush, stirring up dust that sifts through the open windows to cover me with a chalky grit until my bones grate and I think I am dust myself. I work the ridges back and forth, ten, fifteen miles, hunting antelope. It's not that antelope are hard to find. Over every second ridge I see them disappearing, drinking wind at an easy forty-mile-per-hour gambol.

Speed and alertness are the antelopes' only natural defense, though I sometimes think they run for the sheer joy of it. I attempt a surprise attack, driving nonchalantly past a nervous herd then circling around and parking out of sight behind a ridge. I trot down a draw then climb the ridge in a crouch, elbowing my way like a boot-camp trainee through sage and cactus up the last five yards to the ridgetop. I poke my head over, hoping to be invisible to the herd now grazing quietly in the valley below. But they see me, bark a warning, and are gone in a flash. I walk, disheartened, back to the truck.

Driving over another ridge, a bunch of six antelope starts up suddenly and runs parallel to the road. I gun the engine and it turns into a race they cannot resist. They shift into overdrive, zoom ahead and cross the road triumphantly in front of me with a flash of their white tails. I slam on my brakes and let them go. It's a marvel to watch antelope run in earnest, shifting gears from thirty to fifty miles an hour, their necks stretching out and bodies seeming to float level across the sage. Their spindly legs look like they would snap in this rough terrain, but the bone is dense and strong as a steer's.

They are not antelope at all, really, but an unrelated species evolved from the Ice Age; their proper name—pronghorn—never caught on, which is true for other New World species resembling those of the old: buffalo for bison, and elk instead of wapiti. Before the white man's arrival, pronghorns rivaled the bison in sheer numbers, something like 40 million scattered across the western plains. They are ideally suited to this harsh environment, eating whatever forage and browse the season presents, including sagebrush, a winter staple.

Hunting antelope by stealth is a difficult job; when you see them, they've already seen you. Their eyesight is so well developed that they can spot a small object moving four miles away. Sometimes you can corner them at a fence; having evolved in open country, antelope never learned how to jump. Usually they will crawl under the bottom wire, or turn and run a different direction. But I've found that you can make no reliable assumptions about an antelope's behavior. They have a sometimes-fatal curiosity, a trait which Native Americans used to capitalize on, waving flagging from behind a ridge until the animals sauntered up within bow range. I have tried several ploys along this line with mixed results: an elk bugle tooted in their direction proved alarming, sounding too much like their own warning whistle, while lying face up in the sage counting nonexistent clouds got them interested but didn't seem to draw them closer.

With time and a good rifle, antelope hunting is rarely unsuccessful. But in the first license lottery for the area I wanted to hunt I had failed to draw a regular rifle permit, so I applied for one of the "special" licenses that tend to make this sport more sporting, in my case, a ten-day permit to hunt with a black-powder rifle or a handgun. It was an act of desperation, for on the edge of this brutally dry and dusty summer I knew would come an even more brutal winter, cold, deep and empty. A little antelope jerky would help stretch the flour and beans already piled up in my cabin. Like a duel, it had simply come down to a choice of weapons.

When I was younger I used to watch my father shoot his circa-1820 muzzleloader rifle, which rivals a cannon in the roar of its explosion. So I opted for a handgun, borrowing my father's Colt .44 revolver because its barrel was two inches longer than the Smith and Wesson I regularly carry under the seat of my pickup. The gun has an effective range of fifty yards with a stationary target. If you've ever hunted antelope you know you're lucky to get a target under two hundred yards at a dead run. The first few days of hunting with a handgun were a novelty. The last four have begun to wear on me, and in the final, smoke-chastened light of this day I turn my pickup around, defeated, and head for home.

I am up again at dawn, but the weather has turned cold overnight, spawning an uncharacteristic fog so thick that after half an hour driving along a maze of dirt roads I am completely lost. I stop and wait for the sun to burn through, feeling gaunt for lack of breakfast, having already spent nine days of the ten-day season plying these roads in search of that one perfect, unreasonable, fifty-yard standing shot.

Eventually the sky shows itself again, a deep, gritty, cloudless blue, and I choose a road pointed west. The same herds I chased yesterday are back, having run in a wide circle, but I find myself more interested in looking at the country. This is the rim of no man's land, the northern boundary of a thirty-five hundred-square-mile sinkhole of creosote and poison water, hoodoo rock and shifting dunes. Here the Continental Divide splits around the edges of the Great Divide Basin to enclose an area twice the size of Delaware with few roads and no people. I am drawn to the promise of its emptiness, and hang for most of the day on its edges gazing in, guessing at the possibilities.

When the sun begins angling downward, I choose a road towards home that I haven't tried yet. It is less a road than a swath bladed through the sagebrush, as if a pipeline were buried here but there are no warning markers. I hang my head out the window to look at the rocks turned up, a curious collection, so I stop and get out for a closer look, picking up pieces of water-worn quartz. Where there is quartz, there is sometimes gold, and it is gold, after all, that I am usually hunting. Crawling along on my hands and knees I pick up and examine stone after stone, looking for mineralization in the quartz, studying the host rock, guessing at the deposition of this gravel bench. When I finally look up I see, standing fifty yards away, a very puzzled doe antelope.

I look at her awhile, and she looks at me, taking a few tentative steps forward. I stand up and walk back to the truck; as I was hunting gold I did not, of course, bring my gun. She stands waiting like a lamb while

I open the door and buckle on my gun belt. She watches as I walk the twenty yards back to where I was, her eyes opened wide, expectant. I raise the revolver with two hands then draw it down. My uncle, a former Olympic biathlete, taught me how to shoot when I was young. Breathe out, he said, as you pull down on the sights. At the end of the breath, squeeze the trigger gently. The gun roars and the antelope folds like a broken chair, her spine severed where neck meets shoulder. I walk back to the truck for an axe and a knife.

I remember a time when I was younger, twelve or fourteen, hunting alone in the woods with a .22 rifle, looking for supper or I would have to go without. I shot at a squirrel in a pine tree, wounding it, and it clung to a branch until a second shot dropped it, still alive, onto the soft duff beneath the tree. I only had two bullets, so I pushed it with a stick into a small rivulet of runoff water and held it under, my eyes smeared with tears, hands shaking. I can still taste the bitter, smoky meat of that squirrel, cooked on a stick over my campfire. Then, as now, I wish my culture had given me a ritual for death, some blessing chant to restore harmony to the hole of sound that rings in my ears.

Lacking one ritual I must turn to another, slicing along the breastbone and belly, splitting open the ribcage to spill steaming entrails onto clean sand. I accidentally slice my bare arms on her splintered bones, fresh blood mixing with the caked acrid coat of red to my elbows. It will be days before I've washed away all her blood, but I have taken my first step backwards. I lower the tailgate of the pickup and drag her up by the front legs, caught in a kind of macabre dance as I struggle to tip her in. I am a small woman, the antelope's weight and my own close enough to make the contest doubtful. If the light weren't failing I would take the time to quarter her. As it is, I give a final heave and she slides into the truck bed. I close the tailgate and drive away from the red-eyed sun towards home.

"I CAN'T BELIEVE you live here," someone said once, sitting on my front porch. "This is the most desolate place I've ever seen." He had a point, I suppose, and I'll tell you he didn't stay long. It's a hard drive in here on a rutted, dusty road that gets worse as you go along.

Just when the road turns into a Jeep trail you take a left down a steep hill to a questionable bridge over Strawberry Creek that you must cross at your own peril to reach the four cabins on the other side.

The cabins are of various ages and different stages of disrepair. Two are nestled down in the willow bushes of the creek bottom where beaver dams have pooled the green water. These cabins are filled with the debris of neglect—old bed springs, rusted coffee and tobacco cans, shriveled pieces of harness, scattered chunks of collapsing roof—and they lean at forty-five degree angles waiting for the wind to blow them over.

On a bench above (a bench incongruously carved from the hillside and beautifully walled with native rock, and lawned like a resort—said lawn maintained by voraciously marauding cows—with concrete steps leading from lower to upper level) two more cabins sit. One is obviously older than the other, its logs dark and furrowed by weather, its frame sagging slightly and windows boarded over. I use this cabin to store my tools and assorted junk (which is frequently borrowed by the pack rat who lives there, and only forcibly returned).

Because the first three cabins are occupied by one thing or another, I live in the fourth, a one-room, six-windowed, two-doored, twenty-by-thirty-foot log palace. I propped a bed in one corner, an easy chair in the other, a table in the middle, stocked the cupboards and called it home. I found out soon enough this cabin was also occupied, but that story comes later.

Since my palace sits up on the bench, I have a commanding view of the east-west trending valley and the only road in and out of here, a road older than anyone living can remember. For an even better view, I built a pole ladder to climb up on the roof and survey my queendom. It is a silver-green swath of willow bushes, a swag of golden meadow, folded drapes of lead-colored sage, a swatch or two of aspen grove and a ribbon of sandy road.

A mile down the road is the gold-mining ghost town of Lewiston, tumbled buildings and torn earth, a town so decrepit even the ghosts have gone. Across the ridge to the south and out of sight from my rooftop, the Sweetwater River drops into a deep canyon that carries it out of South Pass and on to the dry plains of central Wyoming. It is a river that lives up to its name, clear and cool and consistent.

When the view from the roof feels too confining, I climb the ridge behind the cabins and what I see is sagebrush in every direction, storm-roughened waves of sagebrush that crest in a spatter of dark, fragmented stone. On the northwestern rim of the horizon, mountains—the tail end of the Wind River Range—stop abruptly, definitively, with a downsweep to South Pass and a corresponding change of hue from pine blue to sage gray.

Wyoming's South Pass is not the lowest crossing of this nation's Continental Divide, but it is the most historically significant. While Montana was ridged with row after row of mountains, Colorado was fronted by a wall of Rockies, and New Mexico was striped with deserts between its somber ranges, Wyoming had an open gate. From the Missouri River, up the broad Platte valley and along the looping North Platte to the cool running Sweetwater, over South Pass and down to the Green, across to the Snake and down the Columbia, in five-months time you and your wagon could reach Oregon.

The Oregon Trail across South Pass was discovered accidentally by a lost Scotsman in 1812, blazed by the fur trading caravans of the 1820s and '30s, made famous by the first slow-rolling emigrant trains of the 1840s, ground to dust by gold-chasing Argonauts in the 1850s, sailed over in stagecoaches and pounded by the Pony Express in the 1860s, and was largely abandoned by 1870 when the transcontinental railroad, unbound by the daily ritual of finding grass and water for livestock, was built on a more southerly route. From its beginning the trail would split and regroup, shorten itself and fork as destinations changed and geography unfolded. In the end the trail was like a skein of yarn unraveled by contrary purpose and knitted back together by constraints of terrain.

One of those constraints was South Pass. The Oregon, California and Mormon trails all converge here to cross the Continental Divide, splitting again beyond the far side in directions as diverse as the ideology of their makers. The pass itself was often a disappointment. "If anyone has pictured to himself the South Pass as threading some narrow, winding, difficult rocky mountain gorge," wrote *New York Tribune* Editor Horace Greeley from his stagecoach view in 1859, "he is grievously mistaken." The broad plateau, thirty miles long and half as wide, is bounded on the north by 12,000-foot granite peaks and skirted on the south by ragged buttes of eroding badlands. The road across the imperceptible 7,550-foot summit was smoother and easier, according to one emigrant, than the road leading up to the nation's capitol building in 1849.

South Pass marked the psychological, if not quite physical, half-way point on the road to California and Oregon. Most emigrants noted the crossing in their journals, and many stopped to celebrate the half achievement of their gilded dreams. In three decades, more than 350,000 people crossed South Pass by wagon, as if the entire city of Cincinnati had picked up and moved, twelve plodding miles a day, from one side of the continent to the other. The tracks of those wagons run right through my front yard, pointing like an arrow of faith, westward.

That would probably be sufficient reason for me to live here. History and archeology have been my passion, and reading emigrant journals on the ground they notably crossed has enchanted many hours of my time here. But South Pass gives me another bonus because it was the site of Wyoming's largest gold rush, and gold is my other passion.

I don't know how or when, exactly, I developed an interest in gold, but I have my suspicions. I grew up on the west side of the Wind River Mountains, less than a hundred miles from here, a third generation Wyoming native. Like many pioneer families who have lived on one piece of ground for more than half a century, mine never threw anything away that might conceivably, someday, somehow, be useful for something. (You can actually estimate the tenure of a family in the rural West by the size and variety of their junk pile: old machinery parts, scrap metal, graying lumber, stripped-down vehicles.) The logical extension in my family was to monitor other people's trash, and one of my uncles (I have five on my father's side, which compounded this tendency to the fifth degree) used to take my brothers and me with him to the town dump once a month or so to scavenge. Treasure, after all, is where you find it.

That same uncle mined gold from the beach sands of Australia when I was a child, and brought back an old-fashioned bottle of pink sand laced with gold, which I still have. My family went gold hunting (without luck) several times when I was growing up. And there was something in the sheer volume of bare rock we lived under, the way the mountains towered skyward and the soil never quite covered the rubble. It weighed on our subconscious minds and shaped our stony view of the world. (As proof I present my two brothers, one a geologist and the other a professional rock climber.)

But I have discovered that there is something else at work, regardless of environment, some indefinable tick in a person's psychological profile that makes them, or doesn't make them, susceptible to gold fever. I can usually tell within five minutes if a person is prone to catching the bug.

"What do you do for a living?" they ask.

"Prospector," I say.

"For gold?" they ask, a little light coming on in their eyes.

"Yeah."

"Have you found much?" They lean closer.

"Oh, enough to keep me interested."

"Where?" they ask breathlessly.

That, of course, is a question most prospectors are not inclined to answer. You'll never meet a more tight-lipped crowd than seasoned gold hunters. One may pull you off into a corner to show you his sample bottle of nuggets—gold has a haunting beauty, like a Van Gogh painting, that demands exhibition—but he's unlikely to tell you where the nuggets came from, and certain not to say how much more he actually has rat-holed away. These veterans are also the most skeptical critics of treasure tales, showing polite interest in your stories of potential riches, but not too susceptible to those stampedes of imagination set off in the uninitiated. It's like they've developed antibodies to the virus and only get a mild fever if any. They have moved from the realm of possibility onto the more solid ground of probability with its valleys of doubt.

The truth, and this is as specific as I'm going to be, is that I've already found gold in some thirty streams and gulches throughout western Wyoming, and I haven't yet begun to look hard. The trick isn't finding gold, because it's all over the place, it's finding a lot of gold in one place so the labor of retrieving it is well compensated. Take, for example, a certain mountain near South Pass, a place I call Ghost Dog

Ridge, that geologists estimate contains some $8.5 billion in gold. Claims on that ridge have passed from one hand to another for a hundred years and no one has figured out how to profitably separate that wealth of gold from the 1.6 cubic miles of gravel it is scattered through.

Guide books will tell you that the best place to look for gold is where it's already been found. In some ways this is a discouraging prospect, hunting for what isn't lost, but it's useful advice if you want to see some color in your pan. My best successes have followed a tangent of this rule—look for the least known gold discoveries. Toward this end I have paged through past centuries' newspapers, pored over out-of-print geology reports, followed rumors and hunted up old location notices. This pursuit of "lost" gold mines has added much sport to the game, moving it back from the land of the probable to the realm of the possible, a stampede of the mind.

It's hard to describe what it feels like when you wash out a pan of dirt and find in the last dark spoonful of sand a trail of gold, pure and glowing and beautiful. If you find a twenty-dollar bill caught in brush on the side of the road, you think about what you can exchange it for. If by some quirk you were handed a real silver coin in change, you might wonder how much it was worth to a collector. But gold, in glistening kernels fresh from the earth like an unwrapped gift, you want to keep, and admire, and remember how lucky you felt on the day you found it.

Disappointment is part of the game, for gold hunting is like buying a lottery ticket. You don't really expect to win, but what if you did? The possibility keeps you going. One day in July by Ghost Dog Ridge I was digging in a sandy draw when my shovel hit wood, a hollow, thumping sound. I uncovered the top of a box. "A treasure box," I thought in the unreasonable realm of the possible. I frantically dug around the sides, throwing sand over my shoulder and down my back. Nearby was a pile of old mining machinery from what must have been a substantial operation. "They had to leave in a hurry," I thought, "and buried their proceeds here. One of the miners was drafted into World War I, the other died in a scarlet fever epidemic. Of course they left a treasure map," I reasoned. "But the lone cedar tree was cut down for a fence post, the map became blurry, changed hands, and landmarks never quite look the same so the gold was never recovered and here it still is and it was only by the sheerest luck that I dug in the right spot," I thought excitedly as I scooped out sand.

When I finally freed the treasure chest—pine planks nailed together in a crate like machinery is shipped in—it turned out to be a box and only a box. I moved on to the next dry gulch and bought another lottery ticket.

You would think, growing up in this neighborhood, I would have discovered South Pass sooner, but for various reasons I seem to have taken the long way around. I thought, with some logic, that the gold rush was over here a long time ago. While the central mining districts have been picked clean of the easy gold, I've found many places on the margins and in smaller districts where, because of economic conditions, or remoteness, or lack of water, much gold remains. Staking a gold claim is as easy as setting a post at each of the four corners and at the discovery site, then filing a location form with the county clerk and Bureau of Land Management.

It becomes more complicated when you have to avoid someone else's claim, but the BLM filing fee was recently raised from ten dollars to a hundred, and many claims held by speculators have since been given up and are available for restaking. You are allowed twenty acres per person, up to 160 acres in association with seven other people. Filing on a claim gives you the exclusive right to mine that claim, but not to use it for any other purpose (like building a house). State environmental laws control the size of your mining venture. Any equipment larger than a shovel usually requires permits and reclamation bonds. I staked several claims before I realized the process isn't necessary unless you're serious about mining the ground.

I am not a miner, just a prospector. There are days when I wish for a burro, faithful companion of the old-time prospectors, but a horse and a pickup have to do. When I first hauled my horse out here, a buckskin mare of mustang descent with black, zebra-striped legs, she galloped to the top of the hill and stood looking west towards home. After a week she began looking east to the empty lands where the wild horses run, and she never looked back. We have an uneasy truce between our lonely vigils. She believes in spooks, and snorts at the wind, and runs to meet me on the ridgetop but only reluctantly comes down from her crow's nest view.

I don't know what she sees beyond the emptiness. Gold, it seems, was the only good reason for anyone to settle here. When the gold was gone, the people went too. It takes a certain eye to see the other possibilities of this place, fifty miles from a grocery store, no phone, no electricity, no indoor plumbing. It is not the kind of life that most people would choose, but it suits me, at the moment, better than anything else I can imagine.

When I said I came here the long way round, I meant half a world. Gold fever is not my only malady; I've also suffered from wanderlust. You'd think Wyoming would be big enough for a small girl, but by age twenty-seven I had tromped through forty-five states and more than a dozen foreign countries, changed my occupation every six months, lived in and out of the back of my truck, and like my wild horse, was always looking beyond the horizon. I saw Mexico City and Montreal, Berlin and Budapest, Paris and Prague, Zurich, Venice, Warsaw, Sarajevo.

When I first imagined living here, I was two thousand miles away, working the swing shift in a factory on the East Coast that made armrests for automobiles—ten hours a night standing in one place in front of a high-speed production line, hands flying and eyes blurred as parts rushed past, with the roar of clanking machines and the smell of melted vinyl—I dreamed then about miles of empty sagebrush until my mind went numb. I wanted, finally, to come home.

I arrived here on a cold, overcast day in late May. It was dusk and the wind was blowing. I looked at the skewed cabins that memory had made whole, and the twisted bridge torn off its moorings by flood. The road in had stretched eternally, so much empty space. It was like stepping out of a lighted house into darkness; you think you are confident at first, but soon you must feel your way through the night and your steps begin to falter. I got back in my pickup, turned around and drove away.

Chapter 2

OCTOBER 28. Weather is mostly clear and windy, gusts to 30 mph, lows last night in the 20s. Past full moon. I worked today on a wood-cutting rack without complete success. My dog, a manic Australian Shepherd whose whole life is focused on going some place, went nowhere.

The fall has come and gone stealthily. One day the aspen leaves turned a brilliant orange gold. The next day they were gone. Now the willows are leafless, and the beaver pond is littered with the orange arcs. Why is it that leaves floating on water are so fascinating? It must be the strange juxtaposition of solid on liquid, our own entrancement with the magic, walking on water.

This time, more truly than midwinter, marks the death of the year, after the leaves are gone and before the snow. The drought still holds, a dry cold petrifying the dead grass and crystallizing the pond. The landscape has taken on a coloring that only drought and cold can compose, a mixture of slate and wheat, surreal as a moonscape. To see the bare mountain peaks to the north, defaced months ago of their last patch of snow, is to hear dry stone trickling down. Here in the breaks the willows are burgundy tipped with silver, the aspen bone white.

When I drove away from here in May it took several weeks to gather my courage, to convince myself to walk on down the path I'd chosen. I had driven by the cabins once, years ago, and it is surprising that I remembered them, let alone dreamed of living in them. (The fact that they were among the few buildings with roofs intact within a hundred square miles might have helped my memory.) Research of county land records turned up the owner, and a little friendly persuasion gained me a lease to move in the middle of June.

Between prospecting trips, I spent most of the summer mending fence and getting to know my neighbors—the troll-faced mink living under the bridge, yellow-bellied marmot, a doe mule deer and her spotted fawn, beaver, badger, muskrats, a family of crows and a nest of Swainson's hawks in the aspens up the gulch, two herds of wild horses, two hundred antelope, and an endless supply of flaccid-faced cows.

They were days of blue sky, blazing sun and an occasional cloud—fair weather cumulus. I pieced together the old barbwire fence a strand at a time, and when I hammered staples into the half-rotted posts the wires whined eerily and a crescendo of blackbirds soared up from the fence line.

There was one good rain at the end of July, enough to slough the road for a week, and then more dry, gusting wind. The ground cracked underfoot, spitting dust, but the myriad springs held on and the draws stayed green, closely cropped by the cattle. This used to be sheep country, back in the twenties. The cabins where I live were originally built as a summer sheep camp. There were thousands of sheep out here then, ranging south and east to St. Mary's Peak, spilling into the canyons and draws. It all came to a crashing end after World War II with the invention of synthetic fibers. The price of wool never really recovered.

Now there are cows, Hereford and Angus, loosed on the open range. Except for quarter-section patches here and there, and odd-shaped patented mining claims, the land is all government owned under the jurisdiction of the BLM, Department of the Interior. The rule here is that if you don't want someone else's cows on your land you have to fence them out. I don't have anything personal against cows, except that they tend to be unruly neighbors, and Robert Frost had a valid point about good fences—"Good fences make good neighbors." So I wrestled with the sharp wire, replaced fallen posts and restored my quarter-section of land to a relatively cow-proof and horse-restraining state.

Despite the physical labor, I have to admit that my life here is not full of hardship. I pay twenty dollars a month in rent and have no utility bills. (It's surprising how inconvenient modern conveniences can be when you have to work hard enough to pay for them all.) I haul my water from a spring in the gulch, and heat it on the wood stove. I take showers on the open back step by hanging a solar unit from the roof beam, an often breezy and chilling experience occasionally interrupted by traffic on the main road. But it all feels luxurious because I can do with my time whatever I feel like.

It might seem harder if I wasn't raised to this life. From an early age my father and uncles trained me in the skills of the outdoors—how to fish and hunt, make snares, use an axe, read tracks, build a fire in the snow, pack almost anything on a horse—for which I am relatively grateful. I spent all my summers in the mountain wilderness, scrambling my horse over rocky trails (trails my grandfather blazed), up through timber that sang in the wind like rushing water, up higher to the edge of treeline and on to the top of the world where sky meets mountain in thin air.

I am less at home in the desert, always careful to carry enough water, and a tool kit to repair the pickup, a '63 Ford saved from the scrap yard. (Rebuilding the engine over the course of a month, I was grateful it was made in an era when every part had an obvious purpose.) I have thought often about my childhood out of time, how the skills I learned should be irrelevant to the modern world (how to make, set up, take down and fold an Indian tipi; how to sling barrel stoves on a horse and not lose your top load; how to see the lion track following the deer track and know if the deer knows she is being followed). I ought to be a walking anachronism, and I think this may explain my fascination with the past—history and archeology, other lives out of time.

But the sum total of those skills has given me a sense of self-sufficiency, a desire to be fearless, a craving for adventure that led me here. This may be only a small step backwards in time, but in the quiet, empty days here, it feels big enough.

OCTOBER 30. I looked out the window this morning to see a small herd of wild horses grazing beside the road, all stallions driven away from the main herd of mares and foals by a wild-eyed pinto. They usually hang to the east, beyond Lewiston, where the road is bad enough to keep traffic to a minimum. One afternoon in midsummer on the edge of a thunderstorm I watched this not-too-amiable company of bachelors stage a running fight among themselves, thundering across the sage slashing their hooves at each other, biting necks and falling to their knees to nip at throats. They crashed by me like I wasn't there, rearing and bucking until they spent themselves in a wild run across the breaks. Today they look domestically peaceful crossing the ridge toward the Sweetwater.

I walked up Strawberry Creek this afternoon, curious after reading that five "good-sized" nuggets had been found there. It is a sluggish waterway, with beaver dams every fifty yards and brackish water. Below my cabins the creek sinks into hummocky meadow and the creek bed down at Lewiston is bone dry by midsummer. The gulch beside the cabins was dug up years ago for its gold, and the one time I prospected there, scooping out gravel between two small reservoirs, I found little beads of mercury rolling around in the bottom of my pan like silver bird shot. Mercury is notoriously poisonous—fatal if ingested in large doses, but having a more insidious effect if vapors are inhaled or small amounts are absorbed through the skin over a long period of time—causing teeth to fall out, numbness and tremors in the arms and legs, depression, weight loss, antisocial behavior. In the 1800s, mercuric nitrate was commonly used to make felt for the hat business, thus the phrase "mad as a hatter."

In the gold business, mercury (often called quicksilver, for its color and tendency to divide into meteoric little balls) was an important element in the process of separating gold fragments from other heavy materials, swallowing up precious metals in an amalgam that could be separated later by heating the mixture. As I get my drinking water from the spring that trickles down the gulch, I have since refrained from disturbing any of the gravel, vaguely aware that I might be mad enough.

There is probably mercury in Strawberry Creek too, downstream where it runs by Lewiston's mine shafts and mill sites. Built on top of a gold mine in 1880, Lewiston boasted two general stores, two saloons, one hotel and a dozen residences. The town was not named after Meriwether Lewis, like most western Lewistons, but after Martin Lewis, who happened to build a ten-stamp steam mill there in 1881 for crushing gold ore. The town rode several boom and bust cycles for the next thirty years, and today the only buildings remaining are a half-collapsed log cabin store and livery stable, both piled deep in manure from shade-seeking cows.

Lewiston was an afterthought of the main gold rush to South Pass. Gold may have been discovered here years before the rush to California awakened the nation to the possibility of western treasure. A Georgian working for the American Fur Company found gold on the pass in 1842, but was killed by Indians on his way home to spread the news. This is according to an early newspaper account that does not name its source. Certain parts of the story can be confirmed. The American Fur Company controlled trade from Fort Laramie in the 1840s, and trappers with mining experience from the small gold fields in Georgia and North Carolina occasionally came into the trading posts with gold as well as furs. But if this Georgian was killed, who lived to tell his tale? The story bears a marked resemblance to other tales of lost (and very rich) mines whose finders were either killed by Indians or so demented they couldn't find their way back. Until further evidence surfaces to confirm this story, I will let the jury remain out.

It is certain that "placer" gold from the sands of Strawberry Creek just below my cabins was collected as early as 1860. "*A placer*" is a Spanish phrase meaning "at one's convenience," and placer gold is the flakes, dust and nuggets freed from quartz veins and concentrated by water, the easiest form of gold to mine because weathering and gravity have done the hard work of separating the gold from solid rock. The discovery of gold on Strawberry Creek was resplendent enough to make headlines in the *Denver Daily Evening News* in 1860, but the recipe for a gold rush has complicated ingredients, and it took eight more years to put the Sweetwater mines on the menu.

When the rush came, it wasn't to Strawberry Creek, but to Willow Creek some twenty miles to the northwest, where the town of South Pass City sprang up seemingly overnight, followed by Atlantic City four miles east on Rock Creek, and Hamilton City (later called Miner's Delight) on Spring Gulch. The South Pass rush looked promising enough for the *Chicago Tribune* to send out a newspaper correspondent in 1868 for on-the-spot coverage of this newest strike. By the time he arrived the rocket to riches had already fizzled, the boom had gone temporarily bust, and the correspondent, a displaced Scotsman with a classical education and refined sentiments, spent most of his time wandering around with his hands in his pockets watching the few remaining miners dig for elusive gold. The camps on South Pass would rebound the following year, but in the end there were too many factors working against the miners—isolation, lack of water, and hornet-like Indian raiders—so one by one they moved on.

The rush here to Strawberry Creek in 1880 was more of a trickle, and it came in a time of desperate optimism when mining capital was drying up and every promoter claimed to have the richest property the world had ever seen, occasionally salting their claims with gold from richer mines to bilk the unwary. The mines around Lewiston were worked sporadically for the next half century, but when the easy ore was gone, the town went with it. The buildings, some of which were originally moved in from an abandoned army camp to the north, were hauled away again to a better use in some more hospitable place. No longer are there drunken saloon quarrels, no thundering crash of stamp mills pounding ore, no crunch of shovels or racking of picks, no explosions of dynamite. Only the quiet drip of water in flooded mine shafts and the rush of wind through empty window frames.

South Pass was similar to other gold rushes across the West, donning the gaudy dress of boom times, followed by the tattered rags of bust. Like a vaudeville play that repeated itself for half a century after the discovery of gold "from the grass roots down" in California in 1848, stampeding Forty-niners would perform the dress rehearsal, then the curtain would open in Nevada and Colorado in 1859, Idaho and Montana in the 1860s, South Dakota in the 1870s, the Southwest in the 1880s, and the Klondike in the 1890s, with a curtain call at all of the above caused by rising gold prices in the 1930s.

It was no coincidence that six to ten years passed between each performance. Two essential elements were needed to start a successful gold rush. The first was a surplus of idle or underemployed labor, the result of older mining regions being worked out in the West, or economic depressions causing hardship in the East (there is a surprising correlation between gold rush dates and the dates of cyclical business contractions in the nineteenth century economy). The second necessary element was a spectacular (or spectacularly publicized) gold strike. To join this drama, players often needed both the push of unemployment and the pull of promised wealth.

Once in the play, miners had to write their own script, stealing lines from previous dramas as time went by, or fabricating new ones when the setting demanded it. They made up their own laws, and sought the most expedient ways to enforce those laws. They blazed their own trails to places no sane men wanted to go. They suffered the hardships of pioneers, and they amplified those hardships by their own greed. They were often surprisingly generous to the less fortunate, and heartbreakingly callous when their own success was at stake. They were gamblers, all of them, with luck that soared and plummeted as if their lives were bet on a turn of the card.

I have come to know some of these characters well, reading the journals and letters left behind when the play was over. It is oddly disorienting how time drops away with the intimacy of these accounts, as if they had mailed the letter a month ago and I was the sister who received it. I reflect on their lives, their experiences, because I have nothing better to do, and I see them like snapshots, that quick and candid look of surprise, pinned up in a circle around me. A few of them were obsessed with gold. Others simply got swept into the tide. Some were lucky, others were not, and one or two made their own luck. Like a good fable maker, I have looked for morals in their stories. Their lives were small patterns in a larger tapestry, and the whole cloth runs out of their century and into our own.

OCTOBER 31. The weather can be uncertain here this time of year, at least one good snow to be expected before Halloween. I watch the horizon anxiously for signs of storm, reminded of a mass grave on Rock Creek, four miles up the road, marking an October blizzard more than a century ago. This is one of the stories I have pondered since I moved here, one of the many small dramas of fortune and misfortune that took place on South Pass, the residue of which has almost disappeared without a trace. For once it is a story that has nothing to do with gold, at least on the surface; looking deeper I can see the same golden moral. If greed is an overwhelming desire to make your life better, then the entire story of western expansion—for gold, for land, even for God—is one of greed. And if heroism is trying hard to overcome the odds against you, those same westerners were heros even when they didn't succeed.

For William James, the story began with a promise. James was a farm laborer living in Gloucestershire, England, and times were hard there in the 1850s, with crippling poverty and little hope for relief. He sometimes had trouble feeding his wife and eight children, the last just lately born, named Jane after her mother. James was a conscientious father who wanted better things for his family, and when Mormon missionaries promised a better life in the far-off American West, James became an eager convert.

The Mormons, whose founder Joseph Smith was murdered by a Missouri lynch mob in 1844, had been pushed repeatedly westward by intolerance of their religion, from New York to Ohio, Missouri to Illinois and beyond into wilderness. When they finally settled in the empty valley of the Great Salt Lake in 1847, they called on their flock of "Saints" to "gather to Zion." In what would become Utah they began to build a kingdom, far from the bigoted edge of civilization, carved out of desert—a new life in a new land. They collected their faithful first from the Missouri frontier, then recruited heavily across the Atlantic, gathering some sixteen thousand European converts between 1849 and 1855.

It was to the poor and dispossessed that Mormonism most appealed, offering a new start in this life as well as salvation in the afterlife. But there were many converts, like William James, who lacked the means to join the gathering in Utah, and a Perpetual Emigration Fund was established to help pay the costs of a trans-Atlantic crossing, and the oxen, wagons and supplies to transport the Saints fourteen hundred miles across the American plains. In 1855 alone, $150,000 from the fund was spent to bring in more than eleven hundred poor Saints from Europe. That same year a drought and grasshopper plague in Utah strained resources for the following year's emigration. Mormon leaders were faced with either cutting numbers at a time when more Saints than ever were clamoring to be let in, or finding a cheaper way to transport their converts.

Mormon President Brigham Young had for several years been considering the feasibility of letting the Saints walk across the plains and pull their belongings and provender with them on handcarts. "Fifteen miles a day will bring them through in 70 days," wrote Brigham Young in a letter to his British agent, "and after they get accustomed to it they will travel 20, 25, and even 30 with all ease, and no danger of giving out." And so the experiment in handcart travel was ordained. It would not be easy, the Saints were warned, but the God of Israel would see them through.

William James had faith in this experiment, handed to him like God's own decree. It seemed a golden opportunity, a chance to start over. James didn't know that little Jane would die out on the bleak American plains, that the journey would be harder than promised. He didn't know anything except faith, and it was the kind of faith we build mostly from hope, faith that if we do the best we can then fate or God will smile on us. When James and his family sailed out of Liverpool on May 4, 1856, there were more than seven hundred other converts on board who probably felt the same; if they did as they were told, if they proceeded with faith, they would be justly rewarded.

But even before they reached the Iowa City railhead nearly two months later, fate had begun to work against them. Stormy winter seas had made their ship late setting sail. And though their ocean crossing was uneventful and orderly, they should already have been walking across a sea of grass. At Iowa City, three handcart companies had already set out before them, and Mormon agents were hard pressed to supply this new company. The Saints spent three precious weeks building their own carts out of green lumber, sewing tents, and gathering a meager food supply. Nearly 500 of them marched out of Iowa City in the sweltering heat of July with 120 handcarts, 6 wagons for extra supplies pulled by 3 yoke of oxen each, and a handful of beef and milk cows—1,400 miles to go.

Captained by James G. Willie, a returning missionary, the company crossed the Missouri River in August, and limped into Florence, Nebraska, their last supply outpost before crossing the wilderness plains. Here they stopped a week to repair carts and resupply. Here, also, some of them began to have doubts. John Chislett, a subcaptain in the Willie Company who gives us the only complete account of the journey, admits that "the elders seemed to be divided in their judgment as to the practicability of our reaching Utah in safety at so late a season of the year, and the idea was entertained for a day or two of making our winter quarters on the Elkhorn, ... or some eligible location in Nebraska; but it did not meet with general approval." A meeting of the whole company was called to discuss the matter, but as Chislett describes them, "the emigrants were entirely ignorant of the country and climate—simple, honest, eager to go to 'Zion' at once, and obedient as little children to the 'servants of God.'"

Levi Savage, a subcaptain who had been across this trail before, presented a solitary and adamant objection to going on. "He declared positively," says Chislett, "that to his certain knowledge we could not cross the mountains with a mixed company of aged people, women, and little children, so late in the season without much suffering, sickness, and death."

What Levi Savage lacked was faith, the others thought. "They prophesied in the name of God that we should get through in safety," recalled Chislett, "one elder even declaring that he would guarantee to eat all the snow that fell on us between Florence and Salt Lake City. ... Were we not God's people, and would he not protect us?" It was simply a matter of faith. And so they rattled out of Florence around the 18th of August, now slightly over four hundred in number, an extra hundred-pound sack of flour slung on top of each already overloaded cart. They went up the traditional Mormon road along the north bank of the Platte River. As the air got drier across the plains, the green lumber in their carts shrank, the axles wore through from grinding dust, and the company was forced to stop often to make repairs.

Once the extra flour was consumed they traveled more quickly, but near the Wood River in Nebraska their cattle were stampeded one night by buffalo. They spent three days looking, but came up thirty head short. Chislett described their plight: "We had only about enough oxen left to put one yoke to each wagon; but as they were each loaded with about three thousand pounds of flour, the teams could not of course move them. We then yoked up our beef cattle, milch cows, and, in fact, everything that could bear a yoke—even two-year-old heifers. The stock was wild and could pull but little, and we were unable, with all our stock, to move our loads. As a last resort we again loaded a sack of flour on each cart."

The carts broke down. The company made repairs and struggled on. Farther up the Platte they were passed from the rear by a group of returning missionaries traveling fast in carriages and light wagons. They would secure provisions for the Willie Company at Fort Laramie, they said. They would send relief from Salt Lake when they got there. At Fort Laramie on the first of October the only provision for the Willie Company was a barrel of crackers. They pushed on to the Sweetwater. At Independence Rock, that great "register" in the desert where emigrants before them had carved names and dates (most in July), they found a note saying supplies could not reach them before South Pass.

The nights were bitter cold along the Sweetwater, and the seventeen pounds of clothing and bedding they had been allowed to bring now proved inadequate. They crawled from their tents in the morning more haggard than when they went to bed. Their flour ration, cut repeatedly to stretch supplies, was down to ten ounces per day per person. "Our old and infirm people began to droop," Chislett recalled, "and they no sooner lost spirit and courage than death's stamp could be traced upon their features. Life went out as smoothly as a lamp ceases to burn when the oil is gone. At first the deaths occurred slowly and irregularly, but in a few days at more frequent intervals, until we soon thought it unusual to leave a camp-ground without burying one or more persons."

William James was feeling the strain. He had worked hard, pulling his littlest children on the cart when they could walk no more. Except for losing little Jane, his family had done better than most, for he had brought a shotgun from England and frequently cooked up a rabbit or sage hen with their small ration of flour. But he was tired, and losing strength. Then, on a sixteen-mile forced march across the barren stretch between Ice Slough and Alkali Creek, it began to snow.

If William James had reflected back on the factors that led him to this point, he would have seen that it was partially fate—small parcels of bad luck dealt out randomly—and partially human error. Some of the error was a matter of expediency, some was a matter of faith. At noon, through the swirling snow, a wagon appeared from the west. Help was on the way, the rescuers said, keep moving. The wagon continued on because another handcart company, behind Captain Willie's, was in even worse straits.

They camped late that night sheltered by willows along the Sweetwater, and woke in the morning to find a foot of snow and five of their company dead. Except for a few crackers they were out of food, too weak to push on. Captain Willie and a companion rode ahead to find the rescue teams while the rest waited a day, two days, three. On the evening of the third day the rescue wagons were spotted to the west and wild shouts of joy roused the camp.

The next morning dawned clear and cold, and the small relief party split up, six wagons staying to nurse along the Willie Company while the rest moved down the river to help the company behind. It took two days to move the camp ten miles to the base of Rocky Ridge where they faced one of the worst sections of the whole trail, a long, steep, battering uphill climb to the South Pass plateau.

It began to snow again as they struggled upward the next day. John Chislett had the job of bringing up the rear. "I had not gone far . . . before I overtook a cart that the folks could not pull through the snow, here about knee-deep. I helped them along, and we soon overtook another. By all hands getting to one cart we could travel; so we moved one of the carts a few rods, and then went back and brought up the other. After moving in this way for a while, we overtook other carts at different parts of the hill, until we had six carts, not one of which could be moved by the parties owning it." One of those carts belonged to William James. Chislett found him sitting helplessly by the side of the road, and taking the shotgun from his cart, Chislett tied a small bundle of clothing to one end, laid it across the man's shoulder and started James walking up the road with his fourteen-year-old son beside him.

Chislett found James again later, sitting by the road near where my cabins now stand. It was dark by then and the ox teams in the rear had balked at crossing the ice on Strawberry Creek, so Chislett had gone ahead to get help from the rest of the camp on Rock Creek. "I got him to his feet and had him lean on me, and he walked a little distance, but not very far. I partly dragged, partly carried him a short distance farther, but he was quite helpless, and my strength failed me. Being obliged to leave him to go forward on my own errand, I put down a quilt I had wrapped round me, rolled him in it, and told the little boy to walk up and down by his father, and on no account to sit down, or he would be frozen to death." When horse teams came back to help those in the rear, they found the boy keeping faithful watch over his father, asleep in the quilt. "They lifted him into a wagon," says Chislett, "still alive, but in a kind of stupor. ... His last words were an enquiry as to the safety of his shot-gun."

The Oregon, California and Mormon Trails all come together over South Pass, braided in and out between this short cut and that good camping place. The trail crosses Rock Creek near a canyon sliced through a mountain of dark, twisted rock. There is an eagle's nest at the top of the cliffs, a tangle of pencil-sized branches with odd bits of rabbit bone and fur tufts. In early summer you can look an eagle in her golden eye there, or watch cliff swallows swirl and dive over sparkling water in soft evening light. You can study the stone monument below, with a bronze plaque listing fifteen members of the Willie Handcart Company who died one night in October and were buried there. It doesn't say that their bones were dug up by wolves and scattered before the next year's migration, and it doesn't say that William James, age forty-six, died of faith.

Chapter 3

IT IS NOVEMBER and I have a hard time accounting for the last month, except that there are four cords of dry pine in a pile outside, gleaned from an old timber sale twenty-five miles from here. I finished the wood-cutting rack, a rickety affair that allows me to stack six ten-foot logs at a time to saw into stove lengths. I've grown very attached to my chainsaw, cleaning and sharpening it after every use—an old machine but faithful—and there is a strange kind of joy in the screaming power that rips through dry wood.

Still waiting for the first snowfall, sometimes I am lucid, if not entirely sane. This night is quiet, the wind having died with the rest of the earth, and the stars come to life. How much we miss in our culture by isolating ourselves from the rest of the living world. How much mystery the night sky holds, how much it enlarges the small self.

Darkness has a way of compressing time, of sending us back to that first ring of firelight beyond which danger lurks. We hunch with our spears close to the brightness while the hair stands up on the back of our necks at things unnamed, unseen, prowling the dark behind us. We are creatures of light. Light is safety. Light is comfort. When you learn to walk in darkness alone and unafraid, you have learned to be not quite human.

Sometimes when I am out at night and the kerosene lamp inside the cabin glows through the windows, I look in as if someone else lived there, inside the safety while I was always outside and alone, lurking in the shadows. Isolation has been the hardest thing to deal with here. I have no one to talk to but myself, no company except a stack of books and the vociferous wind.

I am afraid sometimes, I admit, my mind conjuring any number of beasts in the blackness, but the more nights I spend here the less I find to fear. I still feel vulnerable enough in the daylight to be drawn to the window every time I hear a vehicle churning down the main road. I've noted strange passages through that window, facing as it does the old road to Oregon—covered wagons rattling behind mule teams, high-booted Mormons pulling handcarts—as if the wavy glass itself warped time. One morning I looked out to see a mountain man, bearded and buckskinned, riding down my driveway with a loaded packhorse trailing behind. It seemed unfair that I felt compelled to retrieve my revolver from its hiding place, to run a rag over the shiny nickel plating, to flick the cylinder open and count the bullets. The man turned around at my closed gate and I never got to meet him.

I suppose we are all aware, at least subliminally, that we are afraid of our own ignorance. It must be a survival instinct, to retreat from the unfamiliar. Courage is simply a matter of pursuing your fears, of learning your enemy so well he becomes your friend, of walking away from the light.

NOVEMBER 20. I am busy scripting a docudrama in which a young woman is locked in mortal combat with an overwhelming foe. The title is "Mouse Wars" and I would like to cast someone other than myself in the lead role. Scene One opens in the middle of the night with the girl squawking and flailing her arms as she sits upright in bed. The enemy, it seems, has become so bold as to march across her sleeping face. We cannot see this wily opponent, but we can hear his tiny feet scratch a controlled retreat.

Cut to the next morning when our protagonist seeks revenge, baiting three mouse traps with peanut butter. She captures three POWs within minutes, but finds a fourth looking on nonchalantly. What can she do? Out of ammunition, she goes for a bigger gun only to pause at the last minute when she realizes she will literally blow a hole in the floor and let in enemy reinforcements.

Flashback to beginning of conflict when all is calm in Cabinland. Our heroine leaves a five-pound bag of dog food on declared neutral floor only to find it empty upon her return several days later. Canine companion is chagrined but stoically accepts human food instead. New bag of dog food is placed on top of old kerosene refrigerator, out of temptation's way. Girl is awakened in middle of night by a strange noise. Plink, rattle. Plink, rattle, rattle.

Training her Special Forces Headlamp across the room she is stunned to see one little mouse tossing dog food from the top of the refrigerator to an accomplice below. Her illustrious guard dog claims, when summoned, to be of Swiss descent and refuses to get involved. Camera zooms in on bare feet shuffling across the cold wooden floor, hand jerking open the refrigerator door, dog food going inside refrigerator which has not worked in forty years and will not work again except as single mouseproof container in entire cabin.

Flash forward to cupboard scene where in between shredded tinfoil, torn labels, gnawed plastic and tattered paper we gaze into moist black eyes that look up with uncanny trust and innocence. The camera lingers on the large round ears and soft fur, tan along the back and white on the belly. The cute little deer mouse wiggles his whiskers, contemplates awhile, then flattens his plump Houdini body and slips through a crack you couldn't fit a knife blade in.

Cut to final scene where girl calmly pulls a can of parmesan cheese from the cupboard shelf only to realize the top has been chewed open and the perpetrator is still inside wriggling in cheese fluff. Girl runs screaming from cabin, tossing mouse and cheese. Curtain closes and all mice in the audience cheer. Tomorrow I will work on the sequel, "Look Who's Cheering Now," where, fiction or not, the girl really gets her revenge.

NOVEMBER 26. Overcast today, with a hollow, coughing wind. I pile wood in the stoves to temper the chill, and try not to look out at the lifeless landscape, a gray and empty space. Living in a ghost town (or the suburbs of a ghost town, as I like to say), I half expected this place to be haunted. Ghostly apparitions have been reported in South Pass City to my west, and in Atlantic the wind finds dust to stir and hinges to rattle. But I seem to be on neutral ground here—all headstones long crumbled, grievances dispersed, and suffering borne away—as if the bony fingers of sage have scrubbed the air clean of malevolence and charity both. If there are any voices carried on the long marching winds of this place, I think they are not of my race, but of those who followed the buffalo for ten thousand years, and those before the buffalo hunters who lived in a strange world of giant mammoths and miniature camels, dire wolves and saber-toothed cats on the savanna edges of retreating glacial ice.

Those earlier residents have left enough behind to startle us with our own abbreviated tenure. A flake of chert in sagebrush, the delicate, serrated dart point next to a spring. In dry mountain caves, cord woven from bark and baskets from grass. Pictures in unpredictable canyons, hammered into rock—an elk pierced with an arrow, bear and deer, mountain sheep. Man himself, a stick figure, and dream-like man beasts transcribed there on stone like an exorcism of demons.

Most of what we know about these people has been learned from digging up what they left behind: a mammoth kill in the now bone-dry Big Horn Basin to the east, the meat stacked in a winter cache that was never used. A buffalo jump in the Green River Valley to the west, bones piled on bones, the ancient answer to a modern supermarket. We can trace their cultures by the shape of their weapon tips: four-inch long Clovis points used to kill mammoths, the delicately knapped and fluted Folsom points for the giant *bison antiquus*, leaf-shaped PaleoIndian lance points, smaller Archaic dart points, and the diminutive, sharply pointed arrowheads of the Late Prehistoric.

Before the white man began recording the details of native culture, he had already irrevocably changed it. Historically, we came to know the plains Indian as he was in the mid 1800s, a horse-riding, buffalo-chasing whirlwind of savagery, admired and hated in the same breath. Though the Indian had a long tradition of hunting buffalo, he had only acquired horses—an unintended gift of Spanish conquest to the south—in the early 1700s. He had, for some ten thousand years before, hunted buffalo on foot, a cunning stalker, consummate geographer and constant traveler in pursuit of his prey.

Buffalo are wily creatures, and I speak from having spent some time in their close proximity, not wily in their intelligence but in their unpredictability, the way they can spin on you in an instant, how their eyes roll wildly in the side of their shaggy heads as if they still see you rising from beneath a wolf skin with bow strung. They grunt in chorus like distant thunder, like Indian drums, and when they run they rock across their fulcrum shoulders like perpetual motion machines. An Indian on horseback was almost equal to a buffalo. An Indian on foot was one more wolf on the edge of a thundering sea.

Buffalo were not always the main food source for the local Indians. As the climate began warming at the end of the Ice Age, many of the oversized Pleistocene species (and some of the miniatures like horses and camels) died out, probably helped along by newly-arrived human predators. Eventually even buffalo could not survive in large numbers in the seared landscape west of the short-grass plains. Here, in the dry interior basins and mountain foothills, a new subsistence pattern was developed by the natives, one that relied on small game like squirrels and rabbits, with an occasional larger kill of deer or big horn sheep, and a heavy reliance on wild plant foods.

Archeologists sometimes call these people foragers instead of hunters. I like to think of them as opportunists. You look out across this sage country and wonder how anyone could have survived here, a desolate and unnerving proposition. But sagebrush is always deceptive; it makes its own Lilliputian forest, with canopy and underbrush and pint-sized citizens. The landscape is deceptive too, hiding well its draws and springs and secret oases in the hundred-mile view from horizon to horizon.

The natives knew it well—where to dig arrowroot and sego lilies in spring; how gooseberries ripened first down in canyons and later in the mountains; where they could collect yucca seeds and goosefoot, wild rye and prickly pear; how in winter, rosehips remain on bushes buried by snow. They built low shelters over pits dug in the sand because they knew how the wind blows here. They ate grasshoppers and mice and ants and left little of themselves behind—a few charred seeds in a rock-lined pit, some crushed bone, a flat grinding stone, flakes of chert.

They lived on this land for ten millennia, land that is now uninhabited, nearly thirty-two million acres in Wyoming that the government couldn't even give away. Nothing but sagebrush. If we were to measure the success of a culture as we should, by longevity as opposed to material wealth, we would have to ask who had lived the better life here.

That lifeway had changed by the time the first white men crossed the American continent. Horses had made the natives more mobile and thus more independent of their environment. And the horse had arrived at a time when the climate was relatively cooler and wetter, and buffalo were plentiful even in sage country. Many tribes along the Missouri who had been semi-sedentary farmers gave up their corn patches in mounted pursuit of buffalo and a nomadic life farther west. This shifting of traditional territory, probably accelerated by rippling pressure from the white man's advancing frontier, led to conflict and uncertainty where there had once been a fairly stable pattern of land use.

By the first decade of the nineteenth century, members of the well-mounted Crow tribe (recent migrants from the upper Missouri) were terrorizing everyone from the Black Hills on the eastern border of Wyoming to Bear River on the west. The Shoshones, who had traditionally occupied much of that territory, were suffering from a shortage of horses (mostly because the larcenous Crows had taken them), and were barely keeping themselves alive. Yet within half a century the Sioux, pouring out of Minnesota and the Dakotas, would come to dominate the plains east of the Wind River Mountains, pushing the Crows north to Montana and daring the Shoshones to cross South Pass into the Wind River Basin at their own peril.

In addition to mobility, the most significant effect of the horse on native culture was the complexity it allowed, not only the increase of material goods that could be transported, but an increase in the size of the social group. Where hunter-gatherers on foot are most efficient in small family groups, hunters on horseback could support larger, more concentrated numbers of people. The culmination of this trend is best illustrated by the unprecedented gathering in the summer of 1876 of some ten thousand Sioux, Cheyenne and Arapaho along the banks of a creek they called the Greasy Grass, known to George Armstrong Custer and his ill-fated 215-member command as the Little Bighorn.

But that is another story and a different era. Between Lewis and Clark's first probing steps and Custer's crushing defeat lay three-quarters of a century, and half a continent of conflict between lifeways incompatible and irreconcilable. The struggle for control over western lands would turn into all-out war, with both sides suing for occasional peace. Gold would be at the center of this conflict, drawing white men into territory they would otherwise have avoided. The government had a vested interest in the success of gold miners; gold was money, and given the choice between protecting the natives or adding to the money supply, the almighty golden dollar always won.

It is interesting to trace the course of Indian-white relations before and after the discovery of western gold. For the first fifty years of the nineteenth century, the white man traded and traveled and trapped on the natives' forbearance (or occasional lack thereof). First, explorers threaded their way across the plains and through the mountains, followed by fur traders and trappers, then emigrants. All were just passing through, and as tourists they suffered the risks and vagaries inherent in foreign lands.

The most famous sight-seers of the era were, of course, Lewis and Clark, and their expedition overshadowed a second significant tour group, the Astorians. Immortalized by Washington Irving in their own century, the Astorians have largely been forgotten ever since, but the leaders, west-marching Wilson Price Hunt, and east-wandering Robert Stuart, have much to tell us about the occupants of the "empty" half of America.

As the first whites to cross the western continent after Lewis and Clark, the Astorians saw much to cause wonder but little to covet in that seared and broken landscape. The Astorians' view was of a country as pristine as one could be after ten thousand years of occupation: Indian trails worn deep from common passage, antelope alert and grizzly bears elusive, mountain sheep perched pensively on butte tops, and the trampled tracks of a thousand buffalo. In secluded mountain camps, or deep in the basins, or along the falls of rivers, the Astorians met the natives who lived in that country, and the meetings were generally friendly in that first stage of Indian-white relations, when good will and reciprocity reigned.

Under the direction of John Jacob Astor (fur magnate, friend of presidents, and wealthy New York immigrant), Wilson Price Hunt was ordered to lead a party west from St. Louis in 1810 to help establish a trading post on the Pacific coast. A ship rounding the Horn would meet him at the mouth of the Columbia, and together they were to take the fur trade away from British agents in Canada. Hunt intended to follow Lewis and Clark's water route, but native hostility forced him to abandon the Missouri and head overland through South Dakota.

Hunt surprised a band of Cheyennes just east of the Black Hills who happily traded their horses for trinkets (to them, rare and valuable). "The camp was in the middle of the prairie near a little stream," Hunt observed. "The Indians used buffalo chips as fuel. Their tents are made of the skins of this animal well-prepared, carefully sewn together and supported by poles which are joined at the top." Hunt loaded his new horses and pushed on toward the Big Horn Mountains where he met members of the Crow tribe returning from a trading venture with their Hidatsa relatives on the upper Missouri. "These Indians are such good horsemen that they climb and descend the mountains and rocks as though they were galloping in a riding school," Hunt commented. "There was, among others, a child tied to a two-year-old colt. He held the reins in one hand and frequently plied his whip. I inquired his age; they told me that he had seen two winters. He did not talk as yet."

Washington Irving called the Crows the "land pirates of the Rocky Mountains," and Hunt was careful to keep an eye on his horses and his backtrail, but with sixty-five well-equipped men in his army, he was safe from any direct attack. Crossing more mountains, Hunt arrived in the Wind River valley, where Shoshones directed him over the northern pass of the Wind River Mountains (named Union Pass by patriotic Captain William Raynolds in 1860) into the upper Green River valley, across to Jackson's Hole, and over Teton Pass to the Snake River plateau.

The expedition began to unravel in the violent rapids of the Snake when Hunt made the unwise decision to abandon his horses and try to float to Oregon. It was only with help from the natives that Hunt's party, divided and scattered, staggered into Fort Astoria, nineteen months after leaving St. Louis. His route was hardly repeatable, crossing every mountain range in northern Wyoming, and proving no stream short of the Columbia could be navigated.

It was necessary to send a message back east to inform Astor of the company's limited success, and when Hunt declined to retrace his steps, the job of carrying the dispatch fell to twenty-seven-year-old

Robert Stuart. A delicate-chinned, wild-haired Scotsman, Stuart had arrived at the mouth of the Columbia by sailing ship, and had little experience in wilderness travel. A cool head and perseverance were all that would save him as he was bounced like a billiard ball off the varying pillars of native mercy. Unlike Hunt's large, well-armed contingent, Stuart was accompanied by a mere five men in assorted states of mental and physical dilapidation.

As he paddled and portaged up the Columbia for a month, Stuart found the river natives unpredictably hostile, and it required several battles as well as threats and bribery to work their way upriver. The more white men that traveled through their country, the less the natives were willing to grant free passage, and the Columbia was turning into a water highway. (Similar tensions on the upper Missouri had forced Hunt off that river at an unfortunate location; the mountains he climbed over disappeared southward along the Sweetwater and North Platte rivers, as Stuart would discover.)

Stuart left the Columbia before its junction with the Snake, and set off overland on horses he purchased from the Wallawallas. Across the northeast corner of Oregon, over dry hills of sand and brittle clay, between rocks and bluffs, through high prairie and along mountains, Stuart reached the Snake River at the Idaho border. Looping across Idaho for a month and a half, he encountered many small bands of Shoshones fishing for salmon, one of which told Stuart about an easier pass *south* of Hunt's Wind River Mountain crossing that would save much time and distance.

"Learning that this Indian was perfectly acquainted with the route, I without loss of time offered him a Pistol, a Blanket of blue Cloth, an axe, a knife, an awl, a fathom of blue Beads, a looking-glass, and a little Powder & Ball, if he would guide us from this to the other side, which he immediately accepted." Two nights later the Indian guide decamped in the darkness with Stuart's best horse.

Stuart's party managed to guide themselves as far as Bear River, picking up a trapper who with four companions had left Hunt's group the previous fall. The trappers had ranged 450 miles south and east during the winter, only to be robbed twice by Arapahos. "They suffered greatly by hunger, thirst, and fatigue, [and] met us almost in a state of nature, without even a single animal to carry their baggage," Stuart noted, unaware he would shortly find himself in the same condition.

He was on the Wyoming border, within a hundred air miles of South Pass on the twelfth of September when twenty-one overly friendly Crow Indians poured into his camp. "They conducted themselves in such a manner as made it requisite for all of us to keep guard the remainder of the night ... but notwithstanding our vigilance, they stole a Bag containing the greater part of our kitchen furniture."

Terrified the Crows would come back for the group's horses, Stuart fled north, making a wild, 250-mile loop back down the Snake to Hunt's previous route—over Teton Pass, along the Hoback River and into the Green River valley—sixty miles from where he began. The Crows got his horses anyway. Catching up with Stuart in less than a week, they read him a verse from the second chapter of Indian-white relations. Reduced to law, it would sound like this: by native decree, those who were not strong enough to hold on to what they had were divested of it, and if they defended their property too diligently they sometimes also parted with their lives.

Stuart, who wisely didn't resist the Crow raid on his horses, was stoic in his admiration for their boldness, describing how one Indian placed himself in front of the horses while others raised the war whoop, "the most horribly discordant howling imaginable, being in imitation of the different beasts of prey." The horses raised their heads to see what the matter was, while the Indian in front spurred his steed, "and ours, seeing him gallop off in apparent fright, started all in the same direction, as if a legion of infernals were in pursuit of them."

Burning their excess baggage to spite the hovering Crows, Stuart lamented that "we took with us only what we thought absolutely necessary, but we find our bundles very heavy, and the road is by no means a dead level." Set afoot in what he considered a wilderness, and struggling with the tangled geography of disparate drainages, barely a third of the way on his long journey and autumn already begun, Stuart could do nothing but press on toward the sunrise.

In the treacherous mountains of western Wyoming, one member of his group fell ill with fever, while another parted company in a huff only to be found two weeks later, "lying on a parcel of straw, emaciated and worn to a perfect skeleton." A third, and very hungry member (they had been without food for days) suggested drawing lots for a short game of cannibalism. Stuart, himself, found much to reflect on: "former happy days, when troubles, difficulties, and distress were to me only things imaginary. ... If the advocates for the rights of man come here, they can enjoy them, for this is the land of *liberty and equality*, where a man sees, and feels, that he is a man merely, and that he can no longer exist than while he can himself procure the means of [his] support." They finally killed an old bull buffalo in the Green River valley, and ate "voraciously" for half the night.

It was the middle of October before Stuart again reached the vicinity of South Pass where he found forty Shoshones recently victimized by the marauding Crows. For a pistol, an axe, a knife, a tin cup, two awls and a few beads, the Shoshones sold Stuart their last horse. Fresh traces of the Crows on South Pass sent Stuart veering off course once more, away from the easy crossing and out into the Great Divide Basin with its grueling terrain and scarcity of water. He reached the Sweetwater near Devil's Gate, and followed its eastward course to the North Platte, a river no white man had seen or even imagined. Stuart was lost.

Giving up hope of reaching St. Louis that year of 1812, Stuart's group made winter camp on a bend of the Platte near present-day Casper, Wyoming, where timber and game were plentiful. They built an eight by eighteen-foot log cabin roofed with buffalo hides, and settled in for the winter. Considering their pasts—one Kentucky hunter, two French-Canadian voyageurs, a trapper, two disaffected partners of the fragile Pacific Fur Company, and one young Scotsman appointed captain, facing each other, red-eyed, in the dim, smoky, narrow space of that hut—I wonder what it was they had to talk about.

The conversation was interrupted on December 10th by a war party of twenty-three Arapahos (on foot) seeking vengeance on their Crow neighbors for horse thievery. "They all ate voraciously, and departed ... with a great proportion of our best meat," Stuart wrote. Feeling suddenly squeezed by unseen Crows to the north and Arapahos to the south, Stuart's party packed up and headed blindly downriver. They waited out the worst weather and arrived at a surprised St. Louis in

May, eleven months away from Fort Astoria which would become a British prize of war that December, ending John Jacob Astor's dream of a Pacific fur empire.

Robert Stuart never set foot in the West again. History has credited him with the discovery of South Pass and, give or take his digressions, the road to Oregon, but he wouldn't travel that road twice, and much of what he knew about the country would remain dimly lit, a vast, uncertain, tangled heap of landscape. The men who eventually followed Stuart's footsteps and untangled his geography played in a classic drama, dressed in buckskin with a Rocky Mountain backdrop and South Pass at center stage.

The one exploitable resource Robert Stuart had consistently noted in that otherwise dreary wilderness was beaver. London fashion demanded a beaver-felt top hat for every gentleman, and there were beaver in every stream pouring out of the snow-capped peaks. Without steel traps, beaver were hard for the natives to catch and not their usual prey. When beaver pelts failed to come to the white fur traders, the traders came to the pelts by sending their own trappers into the mountains. The road they followed would be Robert Stuart's: up the broad valley of the Platte, along the banks of the Sweetwater and across South Pass to Green River.

It would be a risky job—the climate harsh, terrain rugged, and the natives unreconciled. When trader William Ashley first advertised in St. Louis for "Enterprising Young Men" to go to the mountains in 1822, some of those who signed on would become legends in surviving adversity: Jedediah Smith, first to explore the Great Basin into California and back, who lived through a grizzly attack and innumerable Indian skirmishes only to be killed by Comanches at the age of thirty-two; Jim Bridger, guide, trader and teller of tall tales; David Jackson of Jackson Hole; the Sublette brothers; Indian Agent Thomas (Brokenhand) Fitzpatrick; grizzly-mauled and left-for-dead Hugh Glass who crawled three hundred miles to civilization on his hands and knees.

As Frederick Ruxton wrote from first-hand experience, "the trappers of the Rocky Mountains belong to a 'genus' more approximating the primitive savage than perhaps any other class of civilized man; ... their habits and character assume a most singular cast of simplicity mingled with ferocity." But it was the ability of the mountain men to acquire native skills, to live like Indians and often with Indians that allowed them some degree of survival through twenty years of wandering uncharted country in pursuit of beaver.

By allying themselves with a particular tribe, wintering in tipi lodges and marrying Indian women, they gained security within that tribe's territory, but became enemies with neighboring tribes by association. And if a white man killed an Indian (which happened more and more frequently as the era progressed), he started a blood feud that applied to any other white men in the vicinity, making travel across territories a risky business at best.

And the larcenous law of strength in numbers applied to the mountain men as it had to Stuart. They often lost horses and supplies and the shirts off their backs. (It was Lewis and Clark's protégé, John Colter, who was stripped naked by the Blackfeet and given a running start. Six miles of cactus later he dove into the Jefferson Fork of the Missouri and hid under a raft of driftwood, outraging his foiled pursuers.) The natives were still masters of their domains during the golden years of the Rocky Mountain fur trade in the 1820s and '30s, calling a truce for the annual summer rendezvous when trade caravans arrived to exchange whiskey and trinkets for beaver pelts and buffalo robes. First mule trains, then wagons rumbled up the route that would become the Oregon Trail, a route made possible by the broad plain at the end of the Wind River Mountains called South Pass.

But by the 1840s, the beaver hat had fallen out of fashion and the fur trade had shrunk to a few isolated posts—Fort Laramie, Fort Bridger, Fort Hall. The unemployed mountain men drifted off to meager farms in Oregon or Missouri, or offered themselves as guides to the small wagon trains headed blindly west, inching across the prairies like snails with their houses on their backs. The quest for furs had been a brief and reversible invasion into the great western interior, one the natives might have recovered from were it not for the small fact of a gold-studded sawmill in California.

The diehard mountain men kept their Indian wives and their Indian lives, camping along the emigrant trails and trading briefly rested oxen for "tenderfeet." Peter Decker met such traders at the last crossing of the Sweetwater on South Pass. "These men have habits like Indians," he marveled, "long hair, skin clothing, quick perception & active motions." The year was 1849 and the displaced trappers had already rehearsed, in a strange way, a scene that would be played out again on the same stage: a man knee deep in an icy mountain stream, hammering the stake for a beaver trap/shoveling a scoop of gravel into his gold pan; making up for the hardship and isolation by blowing his pile on a three-day drunk at the fur rendezvous/some tent saloon at a boomtown downriver. There was always more where that came from.

MAN'S FASCINATION WITH GOLD dates back to the Stone Age. Bright pellets plucked gleaming from a stream bed could easily be shaped into a bead or amulet, and with the invention of art, gold became one of the most durable and enduring mediums for expression. Gold has a depth and purity of color that seizes the imagination, a beauty that adds exhilaration to any form. Found nearly pure in nature, it was the first metal to be worked by man. Soft enough to be cut apart with a knife and malleable enough to hammer back together, a single ounce of gold could be drawn into thin wire fifty miles long or pounded into a sheet a hundred feet square. Because of its scarcity, gold has always been more precious than utilitarian, useless really, except to fire all our raw emotions.

You might say a golden harp accompanied the birth of civilization, in the Mesopotamian cradle of what is now Iraq. In one of the world's first cities, built over the mud of the Biblical flood, archeologist Sir Leonard Woolley discovered that goldsmiths had already fully mastered their art more than forty-five hundred years ago. Buried in various royal tombs near the brick wall of the great ziggurat temple were golden daggers and spearheads, engraved bowls and fluted beakers, elaborate headdresses of golden beech and willow leaves, as well as cloaks of beads—cascading gold, carnelian and lapis lazuli. There were gold-decorated chariots and the remains of the animals that pulled them, and golden harps draped with the bones of the gold-crowned harpist. The court attendants of the ancient city of Ur were buried *alive* with the bodies of their kings and queens, as many as seventy-four in one grave.

Woolley suggests they drank a strong narcotic first, and came to their fate willingly, dressed in scarlet robes and decked with all their finery, soothed by the sweet music of a harp on the short journey from one world to another. Woolley's description of one of the ladies-in-waiting adds a strangely human note to this grisly practice. She was found with her silver hair ribbon still in her pocket, "just as she had taken it from her room, done up in a tight coil with the ends brought over to prevent its coming undone. ... Why the owner had not put it on, one could not say; perhaps she was late for the ceremony and had not time to dress properly."

The custom of burying gold with the dead to commemorate a person's social standing and provide for them in the afterlife predates civilization, and the profession of grave robber must be nearly as old. In ancient Egypt, where gold seemed as plentiful as the sand in their streets, mountains of stone were stacked into pyramids to protect the treasures of the Pharaoh's last resting place. But no monolith or false passage or hidden doors deterred the thieves for long. In desperation Egyptian priests began hiding the tombs in limestone caves in the Valley of Kings. These too were discovered and plundered, all except one belonging to a forgotten eighteen-year-old boy king named Tutankhamen.

Excavator Howard Carter described his initial view in the fall of 1922, by flickering candle light through a hole in the tomb's door. "At first I could see nothing ... but presently, as my eyes grew accustomed to the light, details of the room emerged slowly from the mist, strange animals, statues and gold—everywhere the glint of gold." In a room separated from the incredible furnishings of Carter's first glimpse, behind gold-sheeted walls in a stone sarcophagus were three nested coffins, the outer two gold-enameled likenesses of a serene-faced boy and the inner of solid gold requiring eight men to lift it from its three-thousand-year-old resting place.

Gold has consistently been the royal metal of choice partly because of its immortal nature. Silver crumbles, wood rots, bodies decay, but buried gold remains as brilliant and uncorrupt as the day it was entombed. A true phoenix, it rises again and again from the ashes, reformed and revalued. A stamped bar in Fort Knox might once have been a golden bird in the Inca king's treasury; or a Pakistani's necklace, the earring of a Sumerian queen. The love of gold has inspired mankind to the highest acts of art and reverence, and the meanest of murder and destruction.

History is filled with wars of conquest, their motivations complicated but the prize of battle inevitably gold. The Mycenaeans raided Crete, and the kings of Greece besieged Troy. King Solomon's gold was stolen by Egypt then Persia stole it back, collecting golden tribute from most of the Near East. Alexander the Great then plundered the Persian capital, Rome conquered Spain, Mark Antony conspired with Cleopatra, and gold flowed back and forth in trade and tribute, booty and bribes.

By the first millennium gold had shifted from the prerogative of god-kings to the foundation of economies and the fuel of power-seekers. In the Roman Republic, enough gold could buy you the dictatorship, and more gold, the army to keep it. As late as the 1500s Machiavelli commented that "gold alone will not procure good soldiers, but good soldiers will always procure gold."

A desperate shortage of gold helped convince Queen Isabella to back Columbus in his westward search for the treasures of the East. Although Columbus never found Marco Polo's golden Cipangu, Spain would eventually grow decadently rich from the Aztec and Inca treasures.

The northern part of the American continent proved less profitable, though it wasn't for want of trying. "No talk, no hope, no work but to dig [for] gold," wrote a disappointed Jamestown colonist in 1608, when their digging came to nothing. Sir Walter Raleigh instructed the first English colonists on Roanoke Island to look for gold and this they did diligently, without success and to the extreme annoyance of their Indian neighbors. When supply ships returned to the island in 1509, the colonists, numbering more than a hundred and including the first English child born in America, had all disappeared. The only clue left behind was the word "Croatan" carved on a tree, and the mystery of their fate has never been solved.

Colonization would continue in America with less pecuniary motives, and eventually the citizens of the new United States came to realize, slowly, painfully, and despite minor gold rushes to North Carolina and Georgia in the early 1800s, that their nation would be no Eldorado, no Biblical Ophir, no fabulous promised land of gold as thick as the mud in their streets.

IT WAS RAINING that day, the 27th of January, 1848, when James Marshall rode away from the south fork of the American River, one of those bone-soaking cold rains of a California winter. It is difficult to guess the thoughts of the thirty-five-year-old carpenter, a man sunken-eyed and grim-mouthed who rumor said could see the future. Certainly he was agitated when he cornered Johann August Sutter at the New Helvetia fortress and whispered that they must talk behind locked doors. "I could not imagine what he wanted," Sutter recalled. "I knew that he was a very strange man and I took the whole thing as a whim of his ... some secret which he considered important." Deep in an inner office, behind thick adobe walls, Marshall revealed his secret. From his pocket he pulled a dirty white rag and in the rag was a small pile, around two ounces, of kernels and flakes of gold.

At least he thought it was gold. Neither Marshall nor Sutter had seen raw gold before, and Marshall, having pried the flakes out of bedrock cracks in the tail race of a sawmill, had applied what tests he knew to determine if it was real. He pounded it flat with a hammer; gold is malleable—soft and easily shaped—while fool's gold, either pyrite or mica, is brittle and will shatter. He put it in a pot of boiling lye, sure to corrode a lesser metal. He compared it to the one gold coin in possession of his work crew, and he rode away in the pouring rain to tell his partner.

Sutter was skeptical as well, consulting a long article in the *Encyclopedia Americana* and applying more tests. He tried to dissolve the metal in nitric acid without result. He balanced an equal amount of silver coins and gold dust on an apothecary scale, then submerged the scale in water, a test of specific gravity devised by the Greek mathematician Archimedes in the second century B.C. (Archimedes, who was taking a bath while contemplating how to determine if a golden crown made for his king was pure, realized the solution lay in the displacement of water. Jumping out of the tub he ran naked down the street shouting "Eureka!" Greek for "I have found it!") Marshall and Sutter watched the gold sink in the water while the silver rose, and at that early moment their lives began to change, slowly at first, and then steadily, all on a downhill course.

Both Sutter and Marshall were characteristic of the pre–gold-rush population in California, restless eccentrics with a habit of leaving previous failures behind. Sutter, a Swiss German, had fled bankruptcy charges (and a family) in Europe, entangled himself in questionable dealings in the Santa Fe trade, joined a fur caravan to Oregon and arrived in California by sailing ship via Hawaii in 1839. By bluff alone he obtained a huge land grant from the Mexican governor and sailed up the Sacramento to found his very own empire.

By the time he reached California, Sutter had invented a past for himself more noble than true. His father was a Lutheran clergyman, he falsely told people, and he himself had been a captain in the elite Swiss Guard. A stocky man with side whiskers and a pugnacious expression, Sutter had a talent for weaving illusions and then believing in them. Yet he found his life complicated by a desire for admiration that nearly always outstripped his abilities. He proved over and over to be a poor businessman, but he seemed to have charm, pretension and optimism enough to surround himself with an aura of success.

James Marshall had none of Sutter's charisma. A moody, private man, Marshall followed the American custom of looking for greener pastures on the western horizon. The far edge of the continent had been painted as the greenest of pastures by missionaries in Oregon and visionaries in the States, but it was a large risk, leaving everything behind but what you could fit in a wagon, making the hazardous five-month crossing, staking claim to land that did not then legally belong to the U.S. (Britain shared jointly in the Oregon country and California was officially a Mexican province until 1848.) Only a few hundred took that risk in the early 1840s, then a few thousand. James Marshall, who had begun life in New Jersey and had restlessly moved on from Indiana to Illinois and Kansas, found himself in Oregon, still discontented, with California the only destination left.

We could speculate what might have happened if Marshall, whose most endearing quality was a touch of mechanical genius, had not agreed to build, in partnership with Sutter, a sawmill in the California foothills, the result of which was the largest voluntary mass migration in world history. Without a golden catalyst, the Missouri frontier might only have inched its way out onto the tall-grass prairie, leaving the dry interior abandoned to the natives for half a century or more. California would have remained in a pastoral backwash, the small flow of emigration destined for the rich farmlands of Oregon. In Utah, the newly established Mormon colony might have collapsed from isolation. It's possible, even, that the American character might not have developed such an avaricious quality. We believed, before gold taught us otherwise, in a sobering lifetime of hard work with little to show for it.

And John Sutter might have realized his dream—a self-sufficient kingdom with himself as benevolent ruler, admired and respected by all. That is what he really wanted. What he got was something else entirely, for when James Marshall inspected the mill race that day in January, 1848, it would be the last time, in streambeds and dry gulches, on gravel benches and rock ledges all across western America, that someone *wasn't* looking for gold.

For all of its good qualities, its beauty and purity, how it is malleable yet incorruptible, gold seems to bring out the worst qualities in man, the least of which is a sudden inability to keep a secret. Marshall's crew, asked to keep silent about the discovery, continued working on the sawmill, torn between the possibility of golden riches and the certainty of a steady wage. But on Sundays, as employee Henry Bigler described it, "down into the tail race we would go … [where] we could pick and crevice with our jack and butcher knives, and we hardly ever failed to get three to eight dollars and sometimes more." Half an ounce of gold on a Sunday afternoon, worth at today's price over $175. It wasn't long before the dam of silence began to spring leaks.

The first to ride the flood were Sutter's grist mill workers down country, informed by Henry Bigler himself. Then Sutter's entire work force departed for the gold fields, leaving wheat unthreshed and cattle to wander. Rumors trickled out to the coast towns but were met with disbelief. America, after all, was not a gold-rich country. And if there was so much gold, why hadn't it been found earlier in California's long history of occupation? The alcalde of Monterey, Walter Colton, a navy chaplain with a strong desire for order in life, recorded the arrival in that city of an ounce-sized nugget but even so, "doubts still hovered on the minds of the great mass. They could not conceive that such a treasure could have lain there so long undiscovered. The idea seemed to convict them of stupidity."

San Francisco residents were skeptical as well, going on about their sleepy business until Sam Brannan, a Mormon elder with a theatrical flair and a mercantile interest in Sutter's Fort, came careening down the streets with a gold-filled quinine bottle shouting, "Gold from the American

River!" Enough gold in one place can motivate the most unmoving. By the end of May the local *Californian* newspaper (which had to suspend publication for a sudden lack of readers), reported "the whole country from San Francisco to Los Angeles, and from the seashore to the base of the Sierra Nevada, resounds to the sordid cry of gold! GOLD!! GOLD!!! while the field is left half planted, the house half built, and everything neglected but the manufacture of shovels and pickaxes."

By the end of June, 1848, San Francisco was essentially a ghost town. Sailing ships, with their forest of masts, were left bobbing in the harbor, deserted by their crews. "The gold fever has reached every servant in Monterey," complained Reverend Colton in July, "none are to be trusted in their engagement beyond a week, and as for compulsion, it is like attempting to drive fish into a net with the ocean before them." Miners poured in from wherever the news was carried: Mexico and Hawaii, Chile, Australia, China. By fall, Oregon had lost two-thirds of its male population.

James Carson, a soldier who deserted with many of his compatriots, described it this way. "A frenzy seized my soul ... piles of gold rose up before me at every step; castles of marble, dazzling the eye ... thousands of slaves bowing to my beck and call; myriads of fair virgins contending for my love ... in short I had a very violent attack of Gold Fever."

If you were to imagine, like the soldier Carson, an opportunity truly glorious in its possibilities, you could not dream much better than reality in the California summer of 1848. New discoveries on river bars and in dry gulches extended the mining region from the Feather River on the north, 150 miles to the Tuolumne on the south. Erosion made much of the yellow metal available on or near the surface, and if the miners bothered digging, it wasn't more than a few feet deep. An ounce a day was the official average in '48, though individual successes were often counted in pounds.

At Hangtown, James Carson reported "the happiest set of men on earth. Everyone had plenty of dust. From three ounces to five pounds was the income per day to those who would work." The first to mine on the middle and north forks of the American removed in a few days "from five to twenty thousand dollars each, and then left California by the first conveyance." On the Stanislaus a few were making $200 to

$300 a day with only a pick and knife. Carson himself cleared 180 ounces (worth today more than $63,000) in ten days from a creek that would be named after him.

Back in Monterey, home-body Walter Colton watched his neighbors return overloaded with gold: $76,840 worth taken out of the Feather River by a party of seven with thirty Indian laborers in three weeks, over $5,000 by one man on the Yuba in sixty-four days, $4,500 from the North Fork in fifty-seven. One soldier made more during a twenty-day furlough than the government would have paid him for a five-year enlistment. "The gold mines have upset all social and domestic arrangements in Monterey," lamented Colton; "the master has become his own servant, and the servant his own lord. The millionaire is obliged to groom his own horse, and roll his own wheelbarrow."

And they were a ragged lot of rich men, for "clothing was not to be had for love or gold," according to Carson, who described a barely recognizable acquaintance as "bent and filthy . . . his hair hung out of his hat—his chin with beard was black, and his buckskins reached to his knees; an old flannel shirt he wore, which many a bush had tore." Such an abundance of gold and so few available supplies caused a huge spiral of inflation. Refined gold was officially valued at $20 an ounce, with placer selling at around $16, but in the mines it dropped as low as $4 per ounce, with $6 paid in the coast towns. Conversely, shovels began selling for $50 and boots for $150. A horse that cost $20 in '47 sold for $200 in '48, only to be turned loose at his destination because "it was easier to dig out the price of another, than to hunt up the one astray."

Edward Buffum was shocked at the $500 price tag of a log cabin on Weber Creek: "a little box of unhewn logs, about twenty feet long by ten wide." Buffum had been discharged in September, 1848, from the First New York Regiment which had sailed around the Horn at the outbreak of the Mexican War for garrison duty in California. He made his way to the only wharf in San Francisco at the end of October, "armed with a pickaxe, shovel, hoe, and rifle, and accoutered in a red flannel shirt, corduroy pants, and heavy boots."

Buffum encountered returning gold seekers at the mouth of the Sacramento. "Whole launch-loads of miserable victims of fever and ague were daily arriving from the mining region—sallow, weak, emaciated and dispirited—but I had nerved myself for the combat, and doubt not that I would have taken passage when I did and as I did, had the arch-enemy of mankind himself stood helmsman on the little craft that was to bear me to El Dorado."

Arriving at the Yuba River, he "started for the upper diggings to 'see the elephant,'" a phrase that at the beginning of the rush meant simply "to satisfy one's curiosity." By the end it would have an entirely different meaning. Buffum washed out his first bit of gold on the Yuba. "I scraped up with my hand my tin cup full of earth, and washed it in the river. How eagerly I strained my eyes as the earth was washing out, and the bottom of the cup was coming into view! And how delighted, when, on reaching the bottom, I discerned about twenty little golden particles sparkling in the sun's rays, and worth probably about fifty cents."

From the Yuba, Buffum moved to Weber Creek, a tributary of the South Fork of the American River, where he bought the $500 cabin in partnership with nine others. Weber Creek was an area of "dry diggings," where most of the gold was found in ravines draining the hillsides. In the bedrock of one such ravine, Buffum found a long crevice. "It appeared to be filled with a hard, bluish clay and gravel, which I took out with my knife, and there at the bottom, strewn along the whole length of the rock, was bright, yellow gold in little pieces about the size and shape of a grain of barley. … I sat still and looked at it some minutes before I touched it, greedily drinking in the pleasure of gazing upon gold that was in my very grasp, and feeling a sort of independent bravado in allowing it to remain there. When my eyes were sufficiently feasted, I scooped it out with the point of my knife and an iron spoon, and placing it in my pan, ran home with it very much delighted. I weighed it, and found that my first day's labour in the mines had made me thirty-one dollars richer than I was in the morning." Buffum and his partners paid off their cabin in a week, with $500 to spare.

Buffum dashed off to the Middle Fork next, hearing "extravagant reports" that the river was "lined with gold of the finest quality. … The news was too blooming for me to withstand. I threw down my pick-axe, and leaving a half-wrought crevice for some other digger to work out, I packed up and held myself in readiness to proceed at the earliest opportunity." On the Middle Fork, Buffum's party of five washed out two pounds of gold the first day, and Buffum found a pocket of nuggets totaling twenty-eight ounces. Winter rains and a rising river chased them out of their canyon, and on the way back to Weber Creek, Buffum stopped for breakfast at Coloma (the town at Sutter's Mill). For sardines, bread, butter, cheese and two bottles of ale he paid forty-three dollars! "If I ever get out of these hills, and sit and sip my coffee and eat an omelet, at a mere nominal expense, in a marble palace, with a hundred waiters at my back, I shall send back a glance of memory at the breakfast I ate at Coloma sawmill."

Back at the dry diggings, Buffum came down with scurvy. "The exposed and unaccustomed life of two-thirds of the miners, and their entire subsistence upon salt meat, without any mixture of vegetable matter, had produced this disease. ... With only a blanket between myself and the damp, cold earth, and a thin canvas to protect me from the burning sun by day, and the heavy dews by night, I lay day after day enduring the most intense suffering from pain in my limbs, which were now becoming more swollen, and were turning completely black." Buffum was saved when one of his partners found some spilled beans that had sprouted, which he ate boiled along with "a decoction of the bark of the spruce tree."

Buffum gave up mining and returned to San Francisco in the summer of 1849 to begin a career as a newspaper correspondent. He had sentimentally saved his first gold flakes, washed out of Yuba River with a tin cup, and wrapped carefully in a sheet of paper. "But, like much more gold in larger quantities, which it has since been my lot to possess," the fortuneless Buffum wrote in 1850, "it has [all] escaped my grasp, and where it now is Heaven only knows."

Many of his fellow miners would wonder the same, for money easily made was just as easily spent. But the difficulties of that year—the inflation that made gold slip through their fingers, the sickness and hardships—all faded in the brighter glow of golden memory, so that California in 1848 would come to define the shimmering dream for every subsequent gold rush: an abundance of metal easily worked, a relative shortage of miners, an astonishing absence of crime. Mining gold was easier than stealing it that year. As James Carson described it, "heavy bags of gold dust were carelessly left laying in their brush homes; mining tools, though scarce, were left in their places of work for days at a time, and not one theft or robbery was committed."

Like a lottery entered with a pick and a pan, you were guaranteed to win that golden summer of '48. Unless, it seems, your name was Sutter or Marshall. James Marshall watched the completion of his sawmill, ripped the first pile of boards, and was hit by the flood of gold seekers. The sawmill closed down because laborers couldn't be had at any price, not with gold lying around in unimaginable quantities. Miners swarmed over Marshall's land, disregarding his previous claims. And they began to follow him, convinced that Marshall had psychic powers to find the yellow metal. In the leaner years after 1850, he narrowly escaped hanging when he couldn't oblige desperate miners demanding he lead them to gold.

It would almost seem that Marshall and Sutter were cursed with the Midas touch. That long-ago king, granted any wish, chose to have everything he touched turn to gold, a delightful choice until it came time for dinner. James Marshall died in abject poverty, and was buried in a pauper's grave. He had become a bitter and distrustful man. If it were true that he could see the future, as some people claimed, he would have left the gold in the mill race and stayed home that rainy day in January, 1848.

John Sutter fared no better. Since he customarily toasted any visitors to his fort, Sutter was soon overrun with too many visitors to stay sober. Miners slaughtered his cattle, and his fort lost its central importance in Sacramento trade. Creditors, stretched along for years with promises of next year's harvest, began to close in. Overwhelmed, Sutter turned all business over to his son and retreated to a small farm which soon was all he had left. The farm house burned to the ground in 1865, and Sutter eventually left California, his dream trampled and his fortunes scattered. In later years Sutter petitioned Congress for recompense of his losses, personally lobbying his case in Washington. Year after year went by until seventy-seven-year-old Sutter, learning his relief bill had once more failed to pass before Congress adjourned, died alone in a Washington hotel room.

JANUARY 22. These are the elements that mark winter in this high desert: a cold and distant sun, lost in a pale sky, always shadowed by night which falls early and solidly on a place with no street lights to hold it back. Blizzards wail through without warning, climbing on the back of the mountains and howling to be let across, leaving a foot or less of snow that the wind claims as its own. At night the stars shine bleakly; they might be freezing to death. If the wind stops blowing, the silence is unnerving.

I come and go to my cabin as the weather and my cantankerous '68 Polaris snowmachine allow. The cabin is drifted over and cradled by snow eight feet deep. I can walk onto the roof, or off the six-foot stone wall in front, a gentle plane. I once wondered where the snow from the bare ridges went to. Kansas, I thought. Nebraska or South Dakota. But I find a good portion has settled here, seeping through cracks to pile on the floor inside, melting slowly in pools as I load both stoves with coal.

When I leased this place in June from a Lander rancher, having finally gathered my courage, we met in a local coffee shop to talk over the details, and I could tell by his long, sideways looks that he thought I was slightly unhinged. The place wasn't built for winter living—a plank-and-tar-paper roof, windows all along the twenty by thirty-foot log walls, delightfully bright but prone to drafts, fifteen miles to a plowed road. In the fall I rechinked gaps between the logs, tacked plastic over the windows, stacked wood against the west wall and installed an antique parlor stove on one end to balance the Old Majestic cook stove at the other. Between the two I waltz with stacks of wood and buckets of coal and I sleep at night under a down bag, two quilts and a blanket.

In the quiet hours, that brief time when the snow is tinged pink then gray with dusk, I sometimes think of my other neighbors—the human ones. Theodore Roosevelt Hurst over the ridge to the south, and Old Lady Gillespie to the east on Radium Springs. I amuse myself by imagining their lives, piecing together their characters. They are dead, you see. Have been for quite some time. Ted Hurst, better known as the Hermit, wasn't quite right in the head. At least that's what I gather from the stories people tell me. He moved into the country in the early thirties, not far from the Sweetwater River where a very charming lady from Atlantic City says she would occasionally encounter him sniping for gold "buck naked."

I have heard that he took regular counsel from his dead wife. That in winter he would walk the thirty miles to Atlantic City and back in a day. A gentleman who now summers in Atlantic told me he long ago saw Hurst crawling through the snow on his hands and knees. Rushing to the aid, as he thought, of a deathly ill man, he found the Hermit picking up beans scattered from a small hole in the bag on his sled. I have clambered into his mine shaft in Sweetwater Canyon, a masterpiece of cut rock running straight as an arrow toward Lewiston where he intended to come out, four miles away.

He may not have been entirely sane, but I feel a large affection for him. I try to imagine his face, and what I see is thin and drawn, with bloodshot eyes that look farther than the wind.

Mrs. Gillespie is another story. She was an ambitious woman, a dreamer. Radium Springs is a bleak place, without trees or any kind of relief from the rugged sage except the spring itself, which bubbles up from a rock-lined hole, as cold and sweet as any water you will taste. She had a hundred plans for making her fortune from that bleakness, claiming minerals that didn't exist, running a hotel where no one wanted to come, shipping poor ore off to special recovery mills. She tried it seems, much like John Sutter, to surround herself with an aura of importance and success. A man I talked to who knew her in his childhood couldn't quite remember if she died in that place, now a jumble of collapsing buildings on a road that few travel. I suppose I admire her too, when I think about it. It would take courage to land there, and a certain stubborn myopia to sustain such a grand and hollow illusion. May she rest in peace.

Chapter 5

THE SUN WOKE ME UP this morning. Or was I awake and waiting for it, my mind turned on full speed, my life flashing through like a movie I couldn't walk out on? Solitude plays such games.

Outside the wind blows, a constant. I listen to it before I go to sleep, and hear it when I wake up. It rattles everything that's loose—the tattered window plastic, the screen door, the stove pipe. When it stops, momentarily, I sway into the stillness. The geography that made South Pass the gateway to the West also made it an open *wind*ow, a gap in the wall of mountains that funnels the west-rising winds into a constant gale.

I have small jobs to do here, in spite of the wind, like chiseling snow out of drifts to melt for drinking water, a metallic tasting, silty liquid. And wood to split. I spent a disproportionate part of my youth learning the art of laying open dry pine, that and hauling buckets of water forty-five little-girl steps up from lake shore to canvas cook tent. Lacking brawn, I had to discover technique; how, for example, you toss the axe head upward in a curve behind you and only apply force on the way down. The axe should be dull so it splits instead of cuts and you must keep your eye on that spot where a crack nears center so the blade lands exactly there. I wasn't consciously aware that splitting wood required technique until I watched someone try it for the first time. She took one mighty roundhouse swing, missing the block entirely and connecting with her leg. She is still walking only because the axe twisted and caught her broadside. I have my own problems here on days when the wind is so fierce the block blows over before I can get the axe in the air, a frustrating exercise in short-circuitry and exclamation.

Yesterday was relatively calm, so I snowshoed up the gulch, pausing under the hawk's tree to look for sign. Nothing. On to the crow grove, and there on an aspen branch was a great horned owl. With his tufted ears he looked almost like a bobcat crouched and ready to spring, his stripes a camouflage in the fine white branches, were it not for the immense yellow eyes watching me watch him. He flew off in a huff, ruffled and unkempt like he had just got out of bed, his downy plumage silent in flight down to the hawk grove. I walked beneath his perch looking for feathers, for there are few things finer to have than an owl feather.

This place has largely been abandoned by the living, the only contrary evidence a rabbit trail crossing my bridge, and willow sticks peeled by beaver floating in a hole broken in the pond's ice. The cows have gone home, and horses headed south. My resident marmot is sleeping somewhere underground. She used to live under the cabin next door until I moved in and ruined the neighborhood. We had a few midnight encounters where her sharp whistle under my feet sent me starward, but I missed her when she moved to the other side of the creek. The antelope have also moved on, to barer pastures to browse on sagebrush. This season shows the logic of their coloring—the white patches so brilliant in summer make them nearly invisible in winter's quilt of sage and snow.

It seems even the landscape has given in to winter. Things don't look like they should. Whole canyons have moved, and draws disappeared, remolded and filled with snow. If you can say nothing else about the breeze here, you can admit it has artistic flair. A passionate sculptor of snow, the wind. It carves and smoothes and packs and repacks with dedication, with its own vision of the universe as cornice and drift and shimmering polish. The wind is winter's alchemist. Long after nature has finished her other changes—rabbit fur from brown to white, all mutable colors drained and bleached, water stilled, ground frozen—the wind continues its incantations, spinning and boiling and casting spells in snow.

Humankind has had a long-lived fascination with nature's ability to change one thing into another, searching twenty centuries for that one key substance or process (which came to be known among alchemists as the Philosopher's Stone) which would allow man to do what nature did so effortlessly, the way a caterpillar was changed into a butterfly, sunlight into green leaves, ocean basins into mountains. If man could unlock the secret, somehow discover the key to transmutation, he could perform marvelous deeds. He could, for example, change lead—a common, lowly metal—into gold, that most precious and beautiful of nature's gifts.

We know now that the goal of alchemy—aurifaction, or the making of gold—is practically impossible, and it seems absurd to us in the light of modern science that so many "mad wizards" devoted so much time to such an impossible task. Yet the practice of alchemy persisted for two thousand years, passed down through generations in mystical manuscripts filled with strange, symbolic illustrations and convoluted riddles of text which could be deciphered only by adepts after years of study and experimentation.

From China to the Arab world, Egypt to Greece and down to Medieval Europe, alchemy proved irresistible. It was based on the logical assumptions of its time, like Aristotle's theory that the world was constructed from four basic forms of matter: fire, water, earth and air. The reduction of a substance to its primary form and the manipulation of its qualities would allow transmutation. In a laboratory that would be the delight of any mad scientist, filled with mortars and crucibles, furnaces and stills, beakers boiling and cauldrons bubbling, the alchemist powdered and heated, mixed and distilled with many an interesting (and often useless) result.

It was also believed that metals grew inside the earth and changed into one another until they reached maturity as gold. (Exhausted mines were occasionally sealed up in hope that the gold would grow back.) It wasn't unreasonable to conclude that with a little help, a piece of lead could be hurried along to its natural golden adulthood. Gold was the obvious focus of alchemy not only because of its economic value. Gold was the symbol of perfection, the ultimate product of nature, the purest outcome of any progression, materially or spiritually. Gold was also a symbol of immortality, and ancient Chinese alchemists believed this literally, concocting elixirs of powdered gold that killed enough emperors for the art to go out of favor in that region.

Alchemy was half experimental science and half religious mysticism, associating with, and often influencing, contemporary religions from Taoism to Buddhism to Christianity. It was an art that sought inner enlightenment and perfection of the spirit, examining that enigma of one thing which becomes another, and pursuing the essence of life itself. Entwined also with astrology, a "science" as old and persistent as alchemy, each alchemical metal was assigned an influencing heavenly body and corresponding sign. Silver, for example, had the moon, a crescent, and gold was associated with the sun, its symbol a dot within a circle ⊙, looking much like the eye of a blackbird.

While many adepts fully believed in the possibility of transmutation, there were, of course, alchemical practitioners of fraud. Their bag of tricks included crucibles with false bottoms of wax hiding gold beneath, hollow charcoal stirring rods stuffed with gold, palming gold from the sleeve, and coloring alloys to look like gold. Their intention was to produce a small amount of gold so unwise investors would give them money to make more gold. The difference, as you know, between a magician and a charlatan is that with magicians we are fooled willingly.

Alchemists of both sorts often attracted the patronage of kings, which wasn't necessarily a good thing. One renowned practitioner was stolen back and forth between warring rulers, rotting most of his life in castle jails where he was held for "safe keeping." Several were hung, and at least one hung himself. Yet alchemy was never fully discredited until replaced by modern science—even the rational Isaac Newton dabbled in the art.

In an older world, where nature was ruled by gods who could only be influenced with sacrifice and supplication, alchemy was one of the few physical expressions available in man's attempt to control nature, a desire we take for granted now that science has given us so much control we are out of control. And in the end we must concede the point. A particle accelerator at the University of California in Berkeley changed a sample of bismuth into gold in 1980. The cost of this transmutation was $10,000; the gold was worth one billionth of a cent.

The two-thousand-year legacy of alchemy is in some ways a small one—contributing to the apparatus and methodology of chemistry, the symbolism of the occult, the phraseology of mysticism—and in other ways not so small, with the discovery of gunpowder, mineral acids and the distillation of alcohol.

But alchemy has also left us with a kind of mirrored allegory. The reflection looking in is of a mad man stooped and wizened by years in pursuit of something absurdly unachievable. The reflection looking out is our own childlike awe and delight when the thing becomes, like magic, something it was not.

YOU MIGHT CALL IT A kind of alchemy, what happened in the eastern half of America at the end of 1848 when word leaked out from California of a monumental gold discovery—the way doubt shifted into belief and belief to benediction. "An unmitigated humbug in which knaves and fools were the partners," declared the *New York Sun* in its first response to the news.

It looked "marvelously like a speculation to induce a rapid emigration," the *Herald* suggested.

"Such stories are usually magnified and embellished by those who undertake to describe them," scoffed the *St. Louis Republican.*

But President Polk's December confirmation of such an abundance of gold "as would scarcely command belief," followed by the arrival of a tea caddy filled with 230 glittering ounces had all the newspapers singing in a different key and their readers joining the chorus. "Every twentieth person we meet on the street is bound for California," wrote the editor of the *Scientific American.*

"The excitement relative to the gold mines ... continues with unabated fervor," the *New York Herald* reported; "all are rushing head over heels towards the El Dorado on the Pacific—that wonderful California, which sets the public mind almost on the highway to insanity."

It was an ironic alchemy because gold would be not the object transformed, but the transformer, a philosopher's stone in itself. Donald Dale Jackson, who has written one of the best descriptions of the rush to California, characterizes well the early flush of excitement. "Gold was not just the dominant subject of this Christmas season in the East, it was the only subject. Newspapers were filled with it. Ministers preached about it. Teachers discussed it with their classes. Stockbrokers gambled on it. Merchants cashed in on it. It was a tide which engulfed everyone. Nothing remotely like it had ever happened in America."

The tide became more like a tidal wave, gaining height and speed in the shallow waters of truth, propelled by American optimism and grand imagination. One self-proclaimed "expert" declared the gold "absolutely inexhaustible." Naval Agent Thomas Larkin wrote from the mines early in '48, before the easy gold had been skimmed off, that mining "requires no skill. The workman takes any spot of ground or bank he fancies, sticks in his pick or shovel at random, fills his basin, makes for the water, and soon sees the glittering results of his labor." Horace Greeley, in the *New York Tribune*, predicted without basis that California would turn out a "thousand million" dollars in gold within four years. The truth would be closer to four hundred million in a decade.

But the idea of so much gold for so little work was irresistible, and not content with printing the truth, newspapers outdid themselves in embellishment until they believed their own exaggerations. It was good for business, with the sale of wide-brimmed hats and high-topped boots and rough canvas coats, sea biscuits, rifles, pistols and tent cloth, as well as all manner of useless contraptions for getting gold. (One company advertised a special grease you could rub on your body then roll down a hill collecting gold on your descent!)

And it was good, thought Horace Greeley, for America. "We don't see any links missing in the golden chain by which Hope is drawing her thousands of disciples to the new El Dorado, where fortune lies abroad upon the surface of the earth as plentiful as the mud in our streets." It was then, the end of 1848 and the beginning of 1849, that the nation learned a language that would henceforth be associated always with gold—the language of hyperbole.

Eighty thousand would leave for California in the next six months, from the East by ship for the quick dash across Panama or the longer rounding of the Horn; from the Midwest by wagon on the Oregon Trail or along southern routes via Santa Fe or Texas or Mexico. Their fervor had been fanned by the press into spontaneous combustion, and all were convinced of certain success and a quick return.

More who wanted to go lacked finances for the journey, and the initial rush was dominated by young, single men slightly better off than their neighbors. John Banks was the kind of man who would have stayed at home had the promised wealth in California seemed more equivocal. Like everyone else he was swept away by the possibilities, but he wavered at the last moment, torn between devotion to his parents and eight siblings on their farm in Ohio, and the adventure and instant fortune of golden California. In the end he joined a company from Athens County, Ohio, who called themselves the "Buckeye Rovers," and set out overland for California in April of 1849.

The Rovers traveled down the Mississippi to St. Louis and up the Missouri to just short of St. Joseph where they supplied themselves with three wagons, eleven yoke of oxen and the tons of food and equipment necessary for the four-month crossing. They left the Missouri May 10, 1849, heading northwest toward the Platte and the growing swell of wagons like a white rolling river on wheels. A mere four thousand emigrants had crossed the Oregon Trail in 1848; thirty thousand would stampede across the same trail in '49. "The road is dotted by wagons fore and aft as far as the eye can reach, which is no short distance," Banks wrote in his diary May 18.

On the Little Blue River, May 23, Banks noted that "we are [now] in the midst of the Pawnee Nation. They are savage and treacherous." Rumors of Indian attacks haunted the emigrants. They posted guards at night, and kept their guns constantly loaded. The fear of Indians was often more dangerous than the reality of Indians. Banks could not confirm any actual deaths by Indian hands, but he noted a number of men who had accidentally shot themselves with guns loaded for their defense. The Indians on the east side of South Pass were remarkably well behaved in the first decade of the gold rush; those on the west side were more troublesome, as the Buckeye Rovers would find out. The balance tipped the other direction after 1860, with the western tribes subdued and the eastern just beginning to fight for their livelihood.

Total statistics of mortality across the thirty-year history of the Oregon-California Trail are tellingly grim: seventeen graves for every one of the two thousand miles; one fatality for every ten emigrants. They drowned trying to cross rivers, were crushed under wagons, and trampled in stampedes. They murdered each other and were hung for their crimes. Ten percent were killed by Indians, picked off when they wandered from the road. A few gave up and killed themselves. Some, like Willie's Handcart company and the Donner party, were caught by winter snows with inadequate food and shelter.

Most, by far, died of disease, and the worst disease was cholera. Carried from India and Asia to the port towns of the Mississippi and settlements upriver, cholera epidemics were spread by contaminated drinking water, and in 1849 alone, killed ten percent of the population of St. Louis. The trail along the Platte was especially dangerous because emigrants dug shallow wells for drinking water, and the wells were used over and over again without knowledge of contamination. Beyond the Platte, cholera deaths dropped dramatically. A bacterial infection of the intestines, cholera kills by dehydration, often within twenty-four hours, a racking, miserable, sudden, ignominious end of a dream.

By the law of averages, the thirteen-member Buckeye Rovers would be luckier than most; none would die on the trail but they endured their share of hardships. Passing Fort Kearny on the Platte—"about twenty houses made of sod ... neither blockhouse nor palisade. A few soldiers and two or three cannon are all the evidence one has that it is not some outlandish village"—Banks began commenting on the emigrants' inability to get along. "One would think that men who left home together would in this wilderness feel outside pressure sufficient to bind them together, but such is not the case. Today I hear two companies have broken up and, indeed, we have not always the best spirit among ourselves."

If the gold rushers were contentious, they were also unpredictable. Banks noted "an Irish woman bringing hens to California; says they are game." A few days later he met a man returning against the tide, "says he can't go all the way. Has money enough; loves his wife more than gold." A woman and her daughter, traveling alone in the rolling race for gold, caught Banks's attention. "It is said she owns a fine farm in Missouri," he pondered. Banks would find the pair later, on Green River, selling whiskey to a raucous crowd for fifty cents a pint.

Trudging up the Platte, Banks grumbled about whole days of rain and bad roads. When the Buckeye Rovers stopped to rest June 4th at the junction of the north and south forks, six hundred wagons passed them. They forded the Platte, "over a mile wide ... little more than two feet deep," and marched on to Fort Laramie. Built as a fur trading post in 1834, Fort Laramie was purchased by the government in June of 1849 as one of three garrisons to guard the suddenly-crowded Oregon Trail.

"Its appearance is military and neat," Banks wrote, preoccupied with the piles of junk the gold rushers had begun leaving on the trail—broken wagons, extra clothing, hundreds of pounds of flour and bacon, saddles and harness and trunks. "Anything that impedes our progress, not necessary to life, is cast away as worthless."

Leaving the North Platte on June 27, the Buckeye Rovers faced a two-day crossing to the Sweetwater, a figurative mine field of ox-killing alkali pools. They reached the Sweetwater safely and camped near Independence Rock, a granite dome with two decades of emigrant names chiseled and scratched and painted on all sides. Banks climbed the rock but did not leave his own name behind. They next passed Devil's Gate, a notch three hundred feet deep blasted by water through a granite wall. "To stand on the topmost stone looking down into the deep, dark abyss, as I do now, language fails me," Banks wrote.

Four days short of South Pass the Buckeye Rovers had to leave one of three wagons, their oxen failing. "Saw two other wagons cast away," wrote Banks in his diary on July 2, "much property of other kinds, three stoves thrown away and three hundred pounds of bacon in one place. The number of dead oxen is such that should they increase in the same ratio they will create a pestilence." The race to the gold fields was taking its toll.

On the fourth of July, John Banks turned thirty-one but could not pause to celebrate. "Thirty-one years since the light first dawned on these eyes and I was folded in a parent's arms. Now for the first time absent on the return of this day." He crossed South Pass and the dry, three-day pull to Green River. At the Green River ferry, more oxen dropping, Banks wrote, "trouble stares us yet, we hope for the best." Ahead was a corrugation of mountains and valleys, stony roads and sagebrush plains. "This whole region is a miserable, dreary waste. … You seldom see a bird and he can scarcely warble for sadness."

Crossing Idaho, a bit of scandal reached Banks's ears, but provided no levity. It seems a Mrs. Jenkins had fallen in love with a Mr. Lancaster, but their affair was complicated by the presence of Mr. Jenkins. Lancaster persuaded his grown son and a friend to get rid of Mr. Jenkins, offering Mrs. Jenkins's daughter as a reward. Mr. Jenkins was lured from the road, shot in the neck and left for dead. But he wasn't dead, and the next passing wagon train picked him up. The Lancasters were tried and sentenced to return to the States or be shot on sight if they continued to California. Mrs. Jenkins was given back to Mr. Jenkins, but remained "inconsolable for the loss of her lover," who turned out to already have a wife and family back in Illinois. At Soda Springs, Idaho, where water bubbled up from the earth like foamy beer, Banks learned that Mr. Jenkins had died from his wounds.

By August 9th, the Buckeye Rovers had reached the headwaters of the Humboldt River in Nevada and the most difficult part of the journey for man and beast. "Hills of sand, and plains of sand," Banks wrote, little grass, poor water, stifling heat and dust. From the headwaters of the Humboldt to the Sink where the river disappeared into sand, the emigrants were harassed by Indians. The local "Diggers," a tribe of Shoshone stock who eked out a living in the Nevada desert, had found the emigrant traffic a veritable smorgasbord. They picked off stock when the emigrants weren't looking, shooting arrows into oxen or driving them off in the night and feasting on the heaven-sent bounty.

The emigrants were losing enough cattle to starvation and exhaustion. They did not take kindly to the Diggers' depredations and fired on any Indian that showed himself. The Buckeye Rovers lost two oxen in one night. One had been butchered only two hundred yards from the trail; "nearly all his flesh and his skin were carried off," Banks wrote in frustration. "Eight of us armed ourselves, determined to give them a chase and worse if we caught them. … We went some twenty-five miles back in the mountains," finding campsites and trails, but never any Indians.

They gave up their chase and began the difficult climb up the Truckee River to Donner Pass, where they found remains of the unfortunate group who took an ill-advised "short cut" in 1846, and were caught on the eastern side of the Sierras by winter snow. As supplies dwindled, the Donner party chewed on rawhide and shoe leather and finally each other. "The account of their sufferings . . . appears horribly corroborated by the piles of broken bones," Banks reflected. "To me it appeared equaled only by some awful shipwreck."

By the time the Buckeye Rovers dropped down to the California headwaters of the Bear River the middle of September, they had lost eleven of their twenty-two oxen and much of their enthusiasm. "Some think they see the elephant," Banks wrote. "If fatigue, weariness, constant excitement, and awful distress ... make the sight, he is surely here." Travelers by other routes fared little better. In Panama they faced jungles of malaria and cholera. Trying to round the Horn they were shipwrecked or lived on moldy biscuits and stale water. The Mexico crossing was imperiled by bandits, and the Southwestern deserts proved a feverish nightmare. All of the gold rush participants had set out with glowing expectations—to dash to California, make their fortunes and be home in a year. No one warned them that their expectations would crash head on with reality. Setting out to "see the elephant," they had been trampled on the way.

The first Forty-niners to reach California traveled by the Panama route, and among them was Bayard Taylor, a precocious, twenty-four-year-old "travel writer" for Horace Greeley's *New York Tribune.* Hoping to win gold with a pen rather than a shovel, Taylor watched impartially while a party dug for gold in the Mokelumne. "When I first saw the men, carrying heavy stones in the sun, standing nearly waist-deep in water, and grubbing with their hands in the gravel and clay, there seemed to me little virtue in resisting the temptation to gold digging." Taylor changed his mind upon seeing the end result of their labor. "When the shining particles were poured out lavishly from a tin basin, I confess there was a sudden itching in my fingers to seize the heaviest crowbar and the biggest shovel."

Flush times were still going strong during the summer of '49, before the tidal wave of gold rushers washed over the diggings. Taylor found a miner on the Mokelumne who was spending his fortune as fast as he dug it out. Nicknamed "Buckshot," the man was close to fifty years old, with a stout frame and rough clothes. "His face was but slightly wrinkled, and he wore a heavy black beard which grew nearly to his eyes and entirely concealed his mouth. When he removed his worn and dusty felt hat, which was but seldom, his large, square forehead, bald crown and serious gray eyes gave him an appearance of reflective intellect—a promise hardly verified by his conversation."

Taylor found Buckshot's habits even more eccentric than his appearance. "He lived entirely alone, and in a small tent, and seemed rather to shun than court the society of others. His tastes were exceedingly luxurious; he always had the best of everything in the market, regardless of its cost. The finest hams, at a dollar and a half the pound; preserved oysters, corn and peas, at six dollars a canister; onions and potatoes, whenever such articles made their appearance; Chinese sweet meats and dried fruits were all on his table, and his dinner was regularly moistened by a bottle of champagne." The man had reportedly dug out $40,000 and had spent it all. "The rough life of the mountains seemed entirely congenial to his tastes, and he could not have been induced to change it for any other, though less laborious and equally epicurean."

An appetite for luxuries among the newly rich was accompanied by a strong desire for laws to protect their interests. California's government was tentative and incapable of law enforcement (had there been any laws), and the job was left to the miners themselves. "When[ever] a new placer or gulch was discovered, the first thing done was to elect officers and extend the area of order," Taylor observed. "Alcaldes were elected, who decided on all disputes of right or complaints of trespass, and who had power to summon juries for criminal trials. ... The result was, that in a district five hundred miles long, and inhabited by 100,000 people, who had neither government, regular laws, rules, military or civil protection ... there was as much security to life and property as in any part of the Union. ... The capacity of a people for self-government," Taylor concluded, "was never so triumphantly illustrated."

The California model of mining districts set up and regulated by the miners, with laws enforced by majority (or sometimes mob) rule, would be employed everywhere across the mining West. But the balance between law and disorder wobbled precariously whenever the seekers of sudden wealth outnumbered the possessors of sudden wealth. The non-Indian population of California swelled from 14,000 in 1848 to 98,000 by the end of '49, and 250,000 by 1852. While the production of gold also increased, it did not keep pace with the population, and order became harder to maintain, both in the diggings and in the cities that funneled gold away from the diggings.

Mining methods also changed with time and population pressures. Gold washing had begun in California with a wooden bowl or tin basin, and the concept became standardized into a stamped steel pan eighteen inches across and four inches deep with sides sloping at a thirty-seven degree angle. Gold is more than seven times heavier than its quartz-sand host, and swirling water spins off lighter dirt while gold settles to the bottom. Fifty pans washed was considered a good day's labor at the beginning of the gold rush, with an ounce recovered from that labor the minimum required to break even.

Rockers or cradles were an improvement over the gold pan in the amount of material they could process. Approximating the size and shape of a baby's cradle, the box was open on the downslope end, with a sorting screen on top to remove larger stones, and riffle cleats on the bottom to stop the gold. A single miner could work a cradle, but it was more efficient if one shoveled dirt on the screen, another poured water on top and a third rocked the box to agitate the slurry pouring through. Three men could wash about a cubic yard of gravel a day with a cradle, but its best asset was the small amount of water required to run it. Water came dear in the dry diggings, especially in late summer.

Sluice boxes and long toms eventually became the most popular implements for gold washing in California and elsewhere. Their design was simple and construction an easy matter of tacking three boards together into a trough, and adding riffle cleats to the bottom. Segments of the sluice could be pieced together indefinitely downstream, with a steady stream of water directed into the top end, and tailings washing out the bottom. Sluice boxes could be fed at any point, but larger rocks had to be flipped back out with a rake, so the sluice needed constant tending. The long tom solved the rock problem with a screened hopper, but could only be fed at one point. Both sluices and toms were usually "cleaned out" once a week (or once a day if the ground was rich enough) by removing the sand from behind the riffles and carefully finishing the separation in a pan.

As times became leaner after 1849, with more miners grubbing for less accessible gold, the location of mining changed along with the technology. Attention began shifting from the rich river bars, which had already been turned inside out, to the ground underneath rivers which had not yet been touched. Mining the river beds meant moving the rivers in an elaborate construction of dams, ditches and flumes.

Miners had to join together in large companies, and spend months building an alternate path for the water, all in a gamble that might, or might not pay off.

Gold carried in a stream stops only where the current habitually slows: around bends forming gravel bars, in pot holes, cracks or ledges, and behind boulders or other obstructions. The rivers draining the Sierras could rise twenty feet in the rainy season, scouring the bedrock with a vengeance. Miners who worked the broad gravel bars usually succeeded, sometimes spectacularly, while those in the river beds more often failed.

Of all the changes in California mining, John Banks was witness. When he first crossed the Sierras in the fall of 1849, battered and discouraged, Banks was further disheartened when he met a soldier returning to the States. "Nine months' labor in the mines gave him knowledge which to us was very important. Simply this; not quite as flattering as we had heard." Thus warned, Banks bought a cradle for ten dollars and went to work on the Bear River. "Worked three hours, made three dollars," he wrote, unimpressed with his first mining venture. "Faint heart never won fair lady; one man says he made two or three thousand in as many weeks." Banks moved on to the American River where he wrote in December, "this week we feel like being in California; we made over two hundred dollars."

But the rainy season began in December, effectively ending river mining as floods drowned the canyons and snow buried the mountain passes. Banks and several of his Buckeye Rover companions built a cabin at Cold Spring, five miles from Sutter's Mill and forty-five from Sacramento, which Banks found to be a city of six thousand. "Of these, fifty may be women and children. Ten of the habitations are tents of cloth stretched on frames; one is made of sheet iron, another of zinc, the balance poor frames; ... ground rent in some instances as high as the best locations in Philadelphia or New York."

At Cold Spring, Banks began placer mining the ravines above the American River. The "dry diggings" could only be worked in the rainy months when run-off water was available to wash the dirt. Miners stampeded alternately between the river bottoms in the summer and the drier hillsides in the winter. "You constantly see men going in all directions," Banks wrote of the restless surge, "some carrying cradles to a place which others are leaving, perhaps to go where the first left."

There were good weeks for Banks, when a pay streak would light up the bottom of his sluice box, and bad weeks of barren dirt and bad weather. "Monday and Tuesday and Wednesday constant snow, since then rain," he wrote in January. Towns unwisely laid out in the lowlands flooded and flooded again. "Three times this winter Sacramento, like Amsterdam, might be traversed in all directions in boats," Banks commented.

In February Banks traveled to Sacramento to buy mules so he could join the rush back to the mountains in spring. "Despite the flood the place has wonderfully improved. The pasteboard and cloth houses have been nearly all washed from their places, but the number of good frame buildings speaks something like permanent improvement." That was all the good he could say about Sacramento, for on entering the city he became ill with fever and "a desperate chill that made my knees strike and my teeth to chatter. ... To be sick in a tavern away from home and dear friends, surrounded by men whose sole objective is gain, anywhere is miserable, in California it is horrible."

Banks managed to recover from his illness and buy his mules ($175 apiece), but it was March before he set out for the remote Middle Fork of the American River. The mules disappeared the first night, and a worried Banks found them again the next day, three miles back down the road. "This species of property is so uncertain few men expect to see an animal again if a few hours out of sight. Horse thieves are not very plenty, but mule thieves seem almost as numerous as mules. Men are trying to make money by every means."

The trail up the Middle Fork was rugged and treacherous, crossing snow still thirty feet deep, and fording frequently the rapid, swollen river. "The mountains descending and ascending are a mile or upwards in length each and some two thousand feet high. ... While standing on the top of one of these mountains I thought the view one of the wildest and grandest in nature. We who are toiling through the mountains may be sick of such sights, yet artists will delight the world with views taken in this region."

John Banks was one of a multitude heading up various rivers in the spring of 1850. "The rush up this way is tremendous," he wrote in April. "Much gold has been dug in this vicinity and doubtless a vast amount will be taken out this season by crowd after crowd, almost one continuous stream of men." Banks staked a claim on the north fork of Middle Fork in a rugged canyon, and packed in a summer's worth of supplies. "The excitement about finding good diggings is wonderful. Men talk of getting gold by the pound. … All are in wild excitement, or calmly waiting for the water to fall that they may jump and catch the gold."

Waiting for the water to fall without any knowledge of what lay beneath. For three months Banks and nine partners labored on a flume and race to move the river out of its bed so they could mine the gravel that was formerly under water. With the race finally completed on July 14, Banks confided, "now that the time is near when we shall see the result, I feel as if a weighty lawsuit were pending and I am very fearful of the result. My only evidence is circumstantial."

He and his partners (as a representative cross-section, four were from Ohio, two New York, two Missouri, one Illinois, and one Virginia) worked hard to roll boulders up out of the drained bed, sift the gravel and wash the sand. "Ten of us made sixty-three dollars yesterday," Banks bemoaned on August 11, "poor pay in California." Other companies were also disappointed, and all began leaving the canyons to look elsewhere for golden fortunes. "We climb the mountains with a heavy heart and a light purse," Banks finally acceded, "some three or four hundred dollars worse than when the race began."

On August 18, he was "writing once more from the roadside. … We have left our claim like hundreds of others. We meet them from Yuba and Feather, and almost every stream. Misery loves company; we have plenty, but the more the worse prospect. Sometimes I feel almost distracted."

By September, after touring the diggings in other canyons, he concluded that "the amount of labor thrown away on dams and ditches in California would make fifty or one hundred miles of railroad in Ohio." And he found out, to make matters worse, that he should have stayed in his winter camp. "A fool may see the error of the past, but to weigh well the future is a difficult task for any man. In leaving Cold

Spring we felt guided by the best information and, buoyed up by hope, endured many hardships, but we erred. Others came and worked a ravine running in front of our cabin and done well. . . . Fickle fortune, she may smile someday, but a great many almost despair."

Banks's fortunes would rise and sink throughout his three-year mining career, with highs and lows as steep as California's mountain canyons. Luck was the gold rush's elusive companion, and it is symptomatic of human optimism that of the two words describing such fickle fate—*bonanza,* Spanish for fair weather, and *borrasca,* meaning storm—only one would remain in our vocabulary. As Banks described it, "if our prospects brighten we feel cheerful . . . when clouds cross our path, the storm and tempest rage within until the sunshine of hope dispels the gloom."

By the fall of 1850, with thousands more arriving in the gold fields by land and sea, it was gloomily obvious that mining had changed in California since the flush days of '48 when a man could dig out a fortune with a knife blade. Too many miners were pursuing gold that was harder to find, sunk deep at bedrock, hidden in pockets, scattered in dirt twice dug over. What had been true in '48, what lured tens of thousands to California's Eldorado, was suddenly an elusive dream. One man's easy fortune had turned into communal hard labor just to stay alive. "You did not tell us of the thousands that make nothing," Banks complained. "No, but you told us of the big lumps."

Banks moved back to the dry digging southwest of Auburn in October of 1850, building a clapboard house fourteen by sixteen feet in the town of Ophir, "a large place with several hundreds of houses." He set up a boot repair business on Second Street in "Buckeye Row," but there were "five shoemakers and not more than work for one" in the town. By mid December cholera had reached his small town, carried across by the newest wave of emigrants. It was, he reported, "very fatal in this place the past week, one day three died. A young man of New York City was conversing with a friend of mine in the evening; four hours later he was a corpse. The burying ground is up the side of a bleak mountain, a short distance from and in full view of the town. The dead are now committed to the ground in a coffin; formerly a blanket was his only covering."

Plagued by disease and desperate for good diggings, half the population of Ophir had deserted by mid December, taking Banks with them. January 5, 1851: "I am now sitting on the banks of the North Fork in a tent. It is raining and Giles is trying to cook. A queer sight; a fire against a rock, a cake baking, but not very fast as the rain puts a veto on the process. Here we make three dollars per day." January 31: "I have just returned from Yuba [River]. Worked hard one and a half days and made two dollars and fifty cents." February 9: "This week made nearly forty dollars. ... When I reflect on the cause of such vast numbers of men absenting themselves from most of the pleasures of civilized life, that it is all for gold, not gold exactly, but that which it commands, and who can say what it does not make obedient to its will? Monarchs reign by it, men are enslaved for it, men and women marry for it. Yes, more immortal spirits worship thee O Gold. Who knows thy power?"

Constantly on the move, like a mad man pursued by demons, by March Banks had jumped over to Nevada City where the "coyote holes" (shafts dug not in rock but in loose gravel down to bedrock) were seventy feet deep. "Whole hills have been torn so the surface is falling in," Banks noted of this unusual form of mining, half placer, half hard-rock. "Many have sheds over the shafts, so that they may work even in the rain. Two men raise the bucket by means of a windlass."

On April 9, 1851, Banks was on his way back to Ophir. "This day two years from home. Now no better off than then and much older." At the dry diggings miners had dug ditches to carry water to the ravines so mining could continue through the dry months, and Banks joined the off-season pursuit. "Worked hard all day Sunday. Could get the water no other day," Banks wrote unhappily. He usually spent Sunday mornings reading scripture and meditating. In the afternoon he wrote in his diary and hoped for a letter from home, "a drop of water to the parched lips of a shipwrecked mariner." Because ditch water was used and reused on its way downhill until nothing was left, Banks often had to work through the night to get a share of the water.

While water was one problem, fire was another. Nevada City burned to the ground a week before Banks arrived. "Such is the rapidity of things here that several large houses are already up, and in a few days all will be replaced." On May 13 he noted that "the great fire in San

Francisco destroying three-fourths of the city has afforded merchants an excuse to raise the price of goods, some articles twenty-five per cent." Much of the accumulated wealth in California would burn up in catastrophic fires, or wash away in floods. With no insurance, it was bonanza one day and borrasca the next. They would rebuild from the ashes. There was always more where that came from.

Back at Ophir Banks rode his own roller coaster of bonanza and borrasca. June 22: "Much better luck this week than usual. Made $135. We took nearly all of that out in two days. ... I should and do feel grateful for such good fortune." But on August 3 he wrote, "Heat excessive. Made nothing this week though we dug some 200 feet of ravine." By September of 1851 Banks had saved $400, but many of his Buckeye Rover companions had done better in the mountains and some had returned home with modest fortunes. "How hard it is to know what is best here," Banks mused. "Mining requires either great hope or recklessness. The former failed me since the first summer, the latter I hope to be forever preserved from." He wouldn't go home without "making his pile," but like any form of gambling, it is easy to lose the winnings if you put them back into the game.

While luck was an unknown variable in the mines, hard labor was a certainty. The sheer physical exertion of mining was a revelation to all who had set out in expectation of an easy fortune. Their most sophisticated tool was a shovel, their only power, human power. "The abundance of gold has not been as much overrated," complained one Forty-niner, "as the labor of retrieving it has been underrated." Like Banks and his boot repair business, many gold seekers looked for easier ways to make a living. They ran express services to deliver mail to the miners at $2.50 a letter. They sold whiskey for fifty cents a dram and milk at $1.20 per quart, freighted supplies to the mines, sawed logs at $2 a piece or herded mules for $4 a week. One of the Buckeye Rovers went home with $1,800 from cutting hay, another cleared $1,000 at the gaming tables.

Only half the population in California ended up mining the gravel for gold. The other half dreamed up ways to mine the miners. There were saloons at every way station, usually canvas tents with a board across two barrels, and gambling houses in every town. "Perhaps there is no place on earth at the present time containing so much active vice flowing along with an equal amount of energetic virtue. Today I saw a handsome young woman engaged at the card table, drinking, gambling, and swearing. Depravity has not had time to demolish all appearance of amiability. Last year she crossed the plains enjoying the respect and esteem of friends and acquaintances. Very many of these unhappy creatures are kept in gambling houses as *bait*."

Some women found more respectable occupations in the mines, and worked hard for their share of the gold. While actual mining was often too difficult for women, their other employments were almost as grueling: washing and mending clothes for the miners, tending the sick in crude hospitals, cooking, and running boarding houses. "I have made about $18,000 worth of pies," one woman declared, but she had to chop and carry her own wood, bake the pies in an iron skillet over an open fire, and only a third of her income was profit.

Mary Jane Megquier described the labor involved in running a boarding house, a common occupation for married women in the gold fields. "I get up and make the coffee, then I make the biscuit, then I fry the potatoes then broil three pounds of steak, then as much liver. ... After breakfast I bake six loaves of bread ... then four pies, or a pudding, then we have lamb, ... beef, [or] pork, baked, turnips, beets, potatoes, radishes, salad, and that everlasting soup, every day. ... For tea we have hash, cold meat, bread and butter, sauce and some kind of cake and I have cooked every mouthful that has been eaten. ... I make six beds every day and do the washing and ironing. ... If I had not the constitution of six horses I should [have] been dead long ago."

When emigrant Luzena Wilson and her husband arrived in California in September of 1849, after a grueling overland journey, a miner offered her five dollars for the pan of biscuits she had just baked for her family. She looked at him in shock, and he doubled the offer. Luzena dreamed that night "that crowds of bearded miners struck gold and each gave a share to her." The Wilsons sold their oxen and bought into a hotel in Sacramento where Luzena ran the kitchen. After two months they sold out for a $400 profit and speculated in barley. The barley got wet and sprouted in one of Sacramento's many floods, and the Wilsons were broke again.

Luzena moved on to Nevada City, set up her stove under a pine tree, bought two boards for a table and began feeding twenty miners for a dollar a meal. She reinvested her profits and soon had a hotel and a store, but all burned down when Nevada City went up in flames in 1851. Moving on to another valley, Luzena started over again from the back of her wagon. "Housekeeping was not difficult then," she commented wryly about the primitive conditions, "no fussing with servants or house-cleaning, no windows to wash or carpets to take up." Her luck was like everyone else's in California—bonanza or borrasca. The yellow dust unevenly won was poured from hand to hand, and the end result of the whole mad scramble was that almost everyone got a little gold and almost no one held on to a lot of it.

With the high price of supplies, and daily board costing three to five dollars (this in an era when Eastern coal miners were paid a dollar a day), with saloons and gambling houses around every corner, and all grabbing for a fistful of dust, it was hard to hold onto anything. John Banks held on to all he could, for the gold was his ticket home. Banks rarely drank and never gambled. He chided himself for spending two dollars to see a circus of acrobats and contortionists in a leaky tent. He complained about the increasing crime and violence, the senseless deaths from murder and suicide, accidents and illness.

An estimated thirty percent of those who set out after gold died in their quest, ten percent on the journey and the rest in California. Two brothers from Banks's thirteen-member Buckeye Rovers were killed. One died when his gun went off accidentally, the other was killed in a skirmish with increasingly hostile Indians. John Banks died at the age of seventy-seven on a farm in Iowa because one day in 1852 his luck changed. "We took out of the claim this week one hundred ounces. ... Thursday we had nearly twenty-nine ounces, the largest day's work I have ever seen." Within a month he was on a steamer home, several thousand dollars ahead of the game, not as fortunate as some, but luckier than most. A photo of Banks in his later years shows a stern-faced man with Lincoln's cavernous features. He sits stiffly in a chair next to his younger, tight-lipped wife, surrounded by five children who grew up in a tranquil country unmolested by elephants.

Banks left California just in time. The average daily income had dropped from an 1848 high of $20, to $16 in '49, $8 in '50 and $5 in 1852, effectively bringing an end to the wild rush for yellow dust. As an innocent observer plunked down in the middle of the melee, Louise Clappe watched the last, golden glow fade in the summer of 1852 on Indian Bar of the Feather River. She had sailed to San Francisco in 1849 with her physician husband. A refined and literate lady from New England, Louise described herself as a "shivering, frail, home-loving little thistle," and her San Francisco friends were aghast when her husband moved to the northern mining region in 1851, and Louise decided to join him.

"Some [acquaintances] said that I ought to be put into a strait jacket, for I was undoubtedly mad to think of such a thing," Louise recalled. "Some said that I should never get there alive, and if I *did*, would not stay a month; and others sagely observed ... that even if the Indians *did not* kill me, I should expire of *ennui* or the cold before spring. One lady declared in a burst of outraged modesty, that it was absolutely indelicate, to think of living in such a large population of men. ... I laughed merrily at their mournful prognostications, and started gaily for Marysville." The trip, by muleback from Marysville, would be more harrowing than expected. Her husband, generally incompetent, lost the trail twice and they had to camp out in the mountains without blankets or food. (Louise would later divorce the good doctor, an event so common in early California that no one thought it remarkable.)

Arriving at last at her destination, Louise found Indian Bar, deep in a canyon on the upper Feather River, so small it seemed "impossible that the tents and cabins scattered over it can amount to a dozen; there are, however, twenty in all, including those formed of calico shirts and pine boughs." The gravel crescent, formed by a bend in the river, was pitted with deep prospect holes and pocked by dirt piles thrown up, with treacherous paths winding between. Sunshine reached the bottom of the canyon only in summer, and when Louise moved in that October of 1851, she could see light gilding the fir-clad ridges, but would not feel its warmth for many months.

Louise's cabin was made of logs, twenty feet square, and was squeezed in close behind the only hotel on Indian Bar—"a large rag shanty," as Louise described it, with a bar room, dining room, kitchen and bowling alley, the name "Humbolt Hotel" painted crudely in red letters. "The clinking of glasses, and the swaggering air of some of the drinkers, reminds us that it is no place for a lady," Louise concluded. Two steps out the back door she described her small cabin: "it is lined over the top with white cotton cloth, the breadths of which being sewed together only in spots, stretch gracefully apart in many places, giving one a birds-eye view of the shingles above. The sides are hung with a gaudy chintz, … the fireplace is built of stones and mud, … the mantle-piece is formed of a beam of wood, covered with strips of tin procured from cans, upon which still remain in black hieroglyphics, the names of the different eatables they formerly contained."

The single window in the cabin had no glass and the canvas door no latch. Louise's furniture was made up of trunks and packing cases, with a board bench for a sofa, and chairs borrowed from the hotel. "I must mention that the floor is so uneven that no article of furniture gifted with four legs pretends to stand upon but three at once," Louise reported, "so that the chairs, tables, etc., remind you constantly of a dog with a sore foot." She had brought along what she considered the essentials of civilized life: linens, pillows, towels, curtains, a hair mattress and a "handsome carpet." With a few treasured ornaments and perfume bottles, and her candle-box book case graced by Shakespeare and Keats, Louise declared that "I am in reality as thoroughly comfortable here as I could be in the most elegant palace."

As comfortable as she claimed herself to be as the single female resident of the community, Louise found herself snowed in for the winter of 1851 in "a place where there are no newspapers, no churches, lectures, concerts or theaters; no fresh books, no shopping, calling nor gossiping little tea-drinkings; no parties, no balls, no picnics … no promenades, no rides nor drives; no vegetables but potatoes and onions, no milk, no eggs, no nothing." As the "only petticoated astonishment on this Bar," Louise occupied herself with short walks to admire the scenery, or a paddle round the small river pool in a log canoe. She even tried her hand at gold panning. "I wet my feet, tore my dress, spoilt a pair of new gloves, nearly froze my fingers, got an awful headache, took cold and lost a valuable breastpin," the sum total of her labor, $3.25 in gold dust.

The 1850 census recorded the population of women in California at less than eight percent outside of San Francisco, which made Louise the object of much admiration on Indian Bar but did nothing to quell the drinking, gambling and "grotesquely sublime" blasphemy outside her presence. Louise described a drunken two-week party at the Humboldt, to which, as a lady, she was not invited. It began Christmas Eve with an oyster and champagne supper followed by three days of dancing and revelry. "On the fourth day, they got past dancing, and, lying in drunken heaps about the bar-room commenced a most unearthly howling; some barked like dogs, some roared like bulls, and others hissed like serpents and geese. Many were too far gone to imitate anything but their own animalized selves." The saturnalia continued well into the new year, and Louise excused the excess by reason of rainy weather and confinement of the population to "a sandy level, about as large as a poor widow's potato patch."

"You know how proverbially wearing it is to the nerves of manhood, to be entirely without either occupation or amusement," Louise confided. Most of the revelers came by her cabin later and apologized, "looking dreadfully sheepish and subdued." Extreme conditions in the mines tended to exaggerate people's eccentricities, turning character into caricature. And the constantly shifting population swirled anonymity like a cloak over the past. The individuals sharing Louise's small world were a parade of *noms de guerre:* "Paganini Ned," a mulatto whose hair was "frizzled to the most intense degree of corkscrewity," cook and gossip-monger at the Humboldt, who played the violin and believed only a lady could appreciate his culinary talents. "The Squire," who had himself officially appointed justice of the peace while the miners ignored him and carried law to their own ends. ("Were it not for his insufferable laziness and good-nature," Louise commented, "he would have made a most excellent justice of the peace.") Then there was "The General," who was the largest and oldest man on the bar, with a "snow-white beard of such immense dimensions, in both length and thickness, that any elderly Turk would expire with envy at the mere sight of it."

All of Louise's notable neighbors were nearly drowned in the flood of new miners that poured onto Indian Bar in the spring of 1852. The remoteness and difficult terrain of the upper Feather River had slowed development, leaving it as one of the few places not yet turned completely inside out. The invading miners began building flumes to move the river and mine in the bed, as John Banks's failed company had done. They brought with them the desperation and recklessness of losers at a high-stakes poker game. "When I came here, the Humboldt was the only public house on the Bar," Louise commented. "Now there are the 'Oriental,' 'Golden Gate,' 'Don Juan,' and four or five others. … On Sundays, the swearing, drinking, gambling and fighting, which are carried on in some of these houses, are truly horrible."

By August the miners were rioting, and the "Vigilance Committee" organized to control the violence turned out to contain the worst offenders. "In the short space of twenty-four days," wrote Louise in consternation, "we have had murders, fearful accidents, bloody deaths, a mob, whippings, a hanging, an attempt at suicide, and a fatal duel. … I think that I may without vanity affirm, that I have 'seen the elephant.'"

She escaped for three weeks to a calmer civilization in the American Valley, and when she returned to Indian Bar the miners were leaving, their last card played and their chips all gone. "Sometimes … fluming companies are eminently successful," Louise noted, "at others, their operations are a dead failure. … In truth, the whole mining system in California is one great … lottery transaction." It was a lottery lost that summer, for on reaching bedrock in the river bed the miners found no gold. Everyone had gambled on success—the storekeepers and freighters, saloons and boarding houses, doctors and lawyers—extending credit for the promise of paydirt. "Of course the whole world (*our* world), was, to use a phrase much in vogue here, 'dead broke.'"

Of the departing population Louise wrote, "such a batch of woeful faces was never seen before, not the least elongated of which was [my husband's] to whom nearly all companies owed large sums. … There are not twenty men remaining on Indian Bar, although two months ago you could count them up by hundreds." Louise waited in her cabin for packers to come with mules to move her out. "I have been wondering and fretting myself almost into a fever at the dreadful prospect of being compelled to spend the winter here, which, on every account, is undesirable." Her furniture had become packing crates once again, the mud chinking in the fireplace was oozing down with rain, the cloth wall hangings were faded and torn, and the carpet worn out.

With the exodus of the miners Louise had witnessed the end of the gold rush in California. Though mining continued, never again would the odds of the lottery look good enough for all to play. California turned itself to the business of the rest of the world, though a little faster paced and, for a century and a half, with a golden tinge that marked it as a land of opportunity, a land of dreams.

For those who had participated in the rush, it was *the* experience of a lifetime, something they would remember vividly in old age, something which made the rest of their lives seem slow and uneventful. And there were others who had become addicted to the search for gold, who could not put down their picks and pans, whose names would appear as discoverers—in all the great strikes east of the Sierras—as "old Californians." Louise Clappe escaped Indian Bar a day before snow closed the way out, but in the end she was sorry to go. "I like this wild and barbarous life," she wrote; "I leave it with regret."

Chapter 6

MARCH 13. The barometer has dropped six-tenths since yesterday and the wind is wailing—40 mph gusts—after two days of relative quiet. Snow a good possibility.

Yesterday was one of those brilliant days that called for a long walk—a sky full of white, wispy mare's tails and a sun just warm enough for the light wind. I walked up Strawberry Creek, keeping to the snow drifts packed hard as concrete. Melt water was running underneath the snow with a gurgling determination. It seems sky and earth can never agree on when spring should come. I turned south toward the Sweetwater after a mile or so, at a gulch with a small reservoir and aspen grove, climbing the ridge that divides the Bullion Mine at Lewiston from the Burr Mine on the Sweetwater.

Tracks crossed the drifted gulch part way up, yesterday's tracks already half obliterated by the sun, the tracks of a lone running coyote. I thought I heard a coyote's wail last night, abrupt and discordant. They talk a banshee language, atonal, high-pitched, and passionately unearthly. Their fur is the color of winter sage and dry grass, and they materialize and disappear in the brush like phantoms. Sometimes you will see them crossing an open flat, their head and tail lowered in shuffling concentration; it is a miserly world they live in, but it is a world all their own. The coyote is now an icon of the enduring West. Their ability to survive us made them legendary, but it was their voices that made them mythical.

On up the gulch to the ridgetop I came upon a prospect hole blasted into a vein of beautiful smoky quartz. The fist-sized chunks were like tinted glass shattered then partially melted back together. No stain of mineralization, and the hole was half-heartedly shallow. Like any typical mining region, prospect holes outnumber mines here by about twenty to one. They are small craters in the earth, from two to ten feet deep, announcing themselves with mounds of rubble piled around their rims. I usually feel compelled to slide down into every prospect hole I find, to study what it was that interested their makers—a fractured vein of glassy quartz stained red to yellow. Gold doesn't often show itself in ore, but needs an assay, a test by fire that drives off impurities, to reveal its true value. I often pick up pieces from these prospect holes, a collection for future reference, a pile of possibilities.

Walking along the ridge top I could see the remains of the Burr Mine below, a worn out cabin and piles of dump rock. A.T. Burr came down from Montana mines in the 1880s. He asked a fellow prospector where some good placer ground might be, and was pointed toward a shallow gulch draining into the Sweetwater. He had washed out all the good gravel when his Chinese cook found the lode. In the next five years Burr amassed some $8,000 by pounding out the gold-laced quartz in a hand mortar, then he sold the mine and moved on. There was a distinct dichotomy in gold booms between placer and hard rock mining. The initial rush was usually for placer gold, because nature had already done the milling, and all a miner needed was a shovel and a pan. But placer rushes were ephemeral. It required a hard rock mine to keep a town from immediately melting back into brush, and a different kind of mindset to develop a mineral lode. Placer gold was for everyman, while hard rock mines required capital and commitment, a combination most prospectors did without.

Winding my way along the scaly ridge top I came across a curious sight, a tiny hedgehog cactus which had excavated a hole for itself in the shale, a two-inch ball of spines with one little peephole to the sky. I could have stepped on top of it and never known it was there. Hiding from the wind, I bet it was. You can see from here the south end of the Wind River Range, bold foothills of timber running up to the shining peaks. It is a view that a camera can never capture, for the thirty miles between here and there spread another thirty miles to each side and the restless eye grasps it all, while the camera lens must exclude what is peripheral and essential.

My cabins from here look insignificant and unlivable, crouched at the base of a ridge five hundred feet high, a tall wave in this sea of trough and crest. Many of the higher ridges around this country are crowned with stone cairns, intricately pieced obelisks taller than a man that stand darkly against the skyline like sentries, looking down. They were built years ago by sheepherders to mark the vicinity of water, a kind of navigation system, and some of them show true artistry. I came across one by Ghost Dog Ridge—two towers of stacked native rock with an arch connecting them—a near replica of Paris's *Arc de Triomphe.*

There is no shortage of rock to work with here. Like sleeping dragons, the ridges bristle with dark, scaly slabs. The rock is more than three billion years old, some of the oldest in the western U.S., laid down on an underwater shelf at the edge of a micro-continent before any sign of life on this planet.

As much as we think about the past and future, we are creatures of the present, and we must labor mightily to imagine anything but the here and now, to think that once there was no mountain here, that we were underwater, that time for the earth moves on a different clock where our lifetime doesn't count a tick on the second hand. Rocks tell stories without us. It can be a fascinating tale if we try to imagine, for example, how the sediments on that continental shelf were compressed, lifted up, folded, heated, split, and the cracks filled, here and there, with a solution of quartz and gold, gold that waited three billion years to give itself up to us.

I would like, just once, to see the whole geologic story in fast forward, with pieces of continents crashing into each other, and long-vanished seas washing in and rushing out as the craton rippled and split and careened around the planet. I have to rely instead on my imagination and the evidence around me: ridges of rock stacked against the mountains like pages of an opening book. Pale sandstone, dark shale, and gritty limestone from an encroaching sea. A thick layer of dolomite descended from a tranquil ocean and risen now into brilliant white cliffs that catch and hold the last light. Siltstone red as native brick, laid down as river mud in the age of dinosaurs, and eroded into a spectacular crimson canyon that is rimmed by cliffs of pink, cross-bedded sand dunes frozen into solid rock. That is the 500-million-year view on the east side of the mountains.

To the west is the remnant of a once-level basin cut to pieces by erosion. It's like driving from day into night when you cross South Pass, a yin and yang of landscape. When the Wind River Mountains were formed by a thrust fault, more than sixty million years ago during the Laramide Orogeny, the eastern half of the fault line was pushed up and over the west, tilting to a thirty-degree angle the once-level layers of sediment on the east side, and burying them on the west. One side of South Pass is a monochromatic bowl of grand, sweeping vista, the other a variegated staircase of rimrock and dusky canyon.

For fifty million years after the mountains were first raised, erosion spread voluminous skirts of debris into the surrounding basins until the terrain was nearly flat. Rivers flowed without obstacles across a land that was then humid and swampy. (You can find pieces of petrified wood now, out on dry, rocky ground where a tree couldn't be made to grow, strange relics of this other time.) When the entire mountain west was again lifted some two million years ago, erosion began excavating features long buried. Rivers which had been entrenched in their courses for millions of years simply dug deeper, carving through any obstacles they encountered.

It seems a curiosity to us now, the way rivers run right through mountains. (The Wind River, for example, sliced a path through the Owl Creek Mountains like a knife through a block of butter. In an odd geographic misconception, it is the Wind River when it enters the mountains, and the Big Horn River when it emerges.) And it was downright spooky to the wagon-train emigrants of the 1800s how the Sweetwater River blasted a 300-foot-deep notch through a granite wall when it could easily have swung a half mile around on level ground. They called the notch "Devil's Gate" and threw rocks from the top into the boiling rapids below.

The last cast member in this play on landscape was the Ice Age, with at least three major glacial advances in these mountains. Ice half a mile thick plucked and ground at rock until the peaks were like knife blades. The glaciers cut deep, u-shaped valleys, piled up ridges of unsorted rock and boulders, scraped out lake beds and left dams to hold the water. While there are still many remnants of a level landscape here, especially on the west side of the Wind River Mountains where locals call their desert of sculpted butte and arroyo "the Mesa," a Spanish word for table top, it is to ice we owe the drama of these mountains carved out of sky. The glaciers have not disappeared entirely. In fact, the Wind

River Range has the largest area of active glaciers in the lower forty-eight. Some years they lose ground, some years they gain, waiting backstage for a curtain call.

MARCH 14. An amazing squall came up yesterday afternoon, just before dusk. First, wind a solid 30 mph with gusts higher, then snow rattling against the cabin like a thousand needles. I looked out the window and the snow was flying horizontally, driven east by the wind with tremendous force. I ran from stove to stove shoveling in more coal, then to the window to wonder what would become of me. Fortunately it blew through in twenty minutes and I could see a red, smoldering sunset.

This is a disturbing place to be when the wind blows so hard, for I can feel the cabin shift under its force, the stress running through the timbers and up my backbone. I hardly slept last night for the raging wind. It is an irrational uneasiness, in the sense that I cannot calm it with reason. I know the roof won't blow off, but that doesn't help the knot in my ribcage. This morning inside and outside temperatures were equal, my indoor thermometer bottomed out at thirty, a coating of ice on the water bucket.

I have been studying ore samples picked up on my walk the other day. I find them mysterious, like they hold some kind of secret I might learn if I turn them over and over in my hands. "Gold is where you find it," the old prospectors say. And no one, yet, has been able to predict with any reliability where gold will turn up. It was persistence and pure dumb luck that guided the prospectors of the last century. Not knowing where to look, they looked everywhere.

You can narrow the search down to "likely" and "unlikely" places, though. Of the three basic rock types, igneous, sedimentary, and metamorphic, gold is usually found in association with the last named. Igneous rocks come from magma, a molten mush of minerals beneath the crust that forms granite if it cools slowly, and lava rock when quickly cooled. Igneous rock is like an ideally desegregated school; the surrounding population of minerals is evenly and equally represented. Gold doesn't like to hang out in such a democratic environment.

Sedimentary rocks are the lithified leftovers of erosion. Sandstone, shale, limestone, and conglomerates (a mix of cemented pebbles and sand) are examples. The differing hardness within and between layers of sedimentary rock is responsible for many of the spectacular landforms in the American West. Harder caprock resists erosion and forms tall turrets and buttes, unbalanced spires and natural bridges. But sedimentary rocks have usually been reworked and winnowed to the point where any gold they once contained is microscopic and widely scattered.

The third category of rock, metamorphic, is made from either igneous or sedimentary rock whose crystalline structure is changed by heat and pressure. Schist and gneiss are metamorphic forms of igneous rock, marble and quartzite of sedimentary. The forces that transform metamorphic rock are similar to the forces that concentrate gold.

Gold ore is formed when heat and pressure from upwelling magma and circulating water drives off elements like quartz and metals in a super-heated liquid that moves upward towards the earth's crust, filling fractures and seeping into the rock above. The contact zone between the magma and the overlying rock is metamorphosed, half changed into something new, and it is changelings that gold likes best.

Mountains are good places to look for gold for two reasons: mountain-building activity creates the kind of heat and pressure that produces ore, and erosion from the uplift increases the chance that the ore will be uncovered. There might be gold under the flat farmlands of Kansas, but it is buried too deep for anyone to care, at least for now.

Economic deposits of gold are currently found in three geologic contexts. The first is in greenstone belts, a collection of undersea volcanic deposits that were metamorphosed in old mountain-building events, some of the oldest rock left on Earth. The great Kalgoorlie mines of Australia are in greenstone. So are the mines on the Canadian Shield, and in the pharaohs' Egypt, and here on South Pass.

More common in America are gold and silver deposits in association with relatively recent Tertiary (post-dinosaur) volcanic activity. The Homestake Mine, extending more than a mile below the surface of the Black Hills in South Dakota, is one example. The gold is too fine to see, and a ton of ore produces only a small button of metal, but the values have been consistent, and consistency itself is valuable in a gold mine.

The third place to find gold is the paleoplacer, an ancient deposit of placer gold whose source and means of deposition have disappeared. This is the prospector's wild card; it could show up anywhere, turning "unlikely" landscape into one big, sod-covered bank vault.

The Rand mines of South Africa—producing more gold in modern times than any other location—are in a paleoplacer. Originally eroded out of a greenstone belt, the gold was layered with sand and cobbles down the rim of a large basin. Eventually the source eroded away, and sediments and lava flows filled the downwarping basin. Miners have followed the slope of this conglomerate reef more than two miles below the surface of the earth where temperatures exceed 125 degrees.

Klondike miners in the Canadian Yukon looked in vain for the source of their treasure, crawling like marauding ants over the bald knob of King Solomon's Dome. But the original vein was long gone, with only a remnant paleoplacer of white gravel on the benches above Eldorado and Bonanza creeks. Paleoplacers are what made the streams in California so rich. Erosion from the Mother Lode (a 120-mile-long ore-enriched fault zone) had been preserved for millions of years under a cap of lava flows. A renewed uplift of the area caused streams to cut through the lava, reconcentrating the gold from older gravels into rich new deposits. More gold is left in the paleoplacers than was ever taken out of California streams (which amounted to some three thousand tons). It is concealed in the hillsides under soil and brush and lava ridge cap, and you can put it in your "likely" file, but you'll never be able to say where exactly the gold will turn up.

Paleoplacers aren't the only wild cards when it comes to gold. Sometimes a gold-laced quartz vein will turn up in an unexpected place. During the Great Depression, two prospectors in southwestern Colorado were following a trail of placer gold upstream. The trail ended at a sandstone cliff, the most unlikely place they could imagine. But cutting through the sandstone was a vein of quartz carrying gold at eight ounces to the ton. The lucky miners climbed out of the Depression with $150,000.

As an example from my own experience, I was out prospecting in the middle of the desert one day when I came across a small mountain of heaped up granite, pink and granular and highly unlikely in my book.

It was the trailer that tipped me off, a small camp trailer tucked away in an aspen grove, visible only for a short distance along a dirt trail seldom traveled. Like a stalking cat, I watched the trailer for a day, but no one came or went. I tip-toed up to it on a road overgrown with brush and tree branches. The license plate on the trailer was nearly twenty years old, and the camping limit on public land is only fourteen days.

Behind the trailer was an old, hewn-log cabin, and through the age-grimed window I could see mining equipment half covered with a dusty tarp. Up the gulch was a trail of broken, pink granite. I followed the trail, mystified at what they could have been mining, and at the head of the gulch was a vertical mine shaft tracing a vein of yellow quartz in a fault line of the granite. There was a solid-looking wooden ladder descending the shaft, but it was too far down from the lip to reach safely, and though I leaned precariously over the edge, I could not see to the bottom of that dark hole.

Evidence suggested that this was an old mine that had been reopened when the price of gold soared over $800 an ounce. The cabin predated the trailer by at least fifty years, and there were remnants of an old tramway that had carried ore down to flat ground for processing. All was abandoned, and what had happened to the miner I cannot say.

MARCH 16. Only down to 20° this morning, winds calming, barometer dropping, heavy overcast.

I walked down to Lewiston this afternoon—not much to see with snow covering the trenches. I thought I could make out the former mill site, burned beams over a concrete foundation. Many of the mills around here were pulled down for salvage during World War II, when gold mining was illegal (not a strategic metal) and iron was at a premium. There is much detritus left—tin cans, broken glass, discarded machine parts. That, after all, is what the historian collects, what is left over when the principles have vanished. We sift through the refuse and try to extract what is meaningful. I find I take an archeologist's delight in trash, for each worthless piece is part of a puzzle made up of other lives and other times. And it is through such context that we make any sense of ourselves.

The West is littered with ghost towns like Lewiston, relics of a brief flirtation with greatness. Abandoned, or nearly so, they are monuments to misplaced optimism, casualties of a limited resource. After the gold was gone they had little to recommend them, propped, as they were, against steep hillsides or sunk in obscure canyons or perched on a mountain, short on water or long on snow. (Phoebe Gustin remembered winters in Lewiston in the 1890s when the blizzards were so bad they had to run a rope from the house to the barn and outhouse so they wouldn't lose their way. Snow covered their house to the roof and they had to tunnel out.)

A few of the early gold towns survived and prospered, like Helena and Boise and Denver, because they were favorably located, and other cities like San Francisco and Seattle were built up by supplying the gold rushes. But most of the gold camps are decaying, board by brick, mere skeletons of the original with bones picked clean. Those still standing are often held up by tourism, their quaint, nineteenth-century frontier architecture preserved like an old photograph. Places like Virginia City, Montana, where among the museum's clutter of relics you can see the gruesome contents of a nineteenth century doctor's bag, or gaze at solemn, yellowed portraits on the walls, of people you don't know who by some trick of the photographer always seem to be looking past and beyond you.

Or Virginia City, Nevada, where you can play the slots at the Bucket of Blood Saloon, or take a shot at the town's most famous one-time resident. He stands there, a caricature in a shooting gallery, in a white suit with the bristling white hair and bushy moustache as we would know him in his famous later life, book in hand and about as far from the Mississippi River as he could get.

IT WAS JULY in the year 1861 that Sam Clemens boarded a stagecoach in St. Joseph to leave the sweltering eastern half of America for the parched and primitive West. He was bound for Nevada with his brother, Orion Clemens, who had been appointed secretary to that new territory by dint of friendship with a cabinet member in President Lincoln's White House. Sam was going to be secretary to the Secretary, a payless job, but he'd always wanted to go west.

Sam and Orion sailed out onto the Kansas prairie in "a great swinging and swaying stage, of the most sumptuous description—an imposing cradle on wheels ... drawn by six handsome horses," along a trail beaten down by 250,000 people in the decade following the discovery of gold in California, a trail that was now a road with way stations every ten miles and a daily coach that could cover the ground in eighteen days that the wagon trains took five months to traverse.

Sam and "the Secretary" were joined by a third passenger named Bemis who carried a six-*barreled* Allen's revolver, one of the marvels of American firearms invention. "To aim along the turning barrel and hit the thing aimed at was a feat which was probably never done. ... Sometimes all its six barrels would go off at once, and then there was no safe place in all the region roundabout but behind it," Sam described. They were accompanied by twenty-seven hundred pounds of mail stacked inside and on top of the coach, which made a very nice bed when the bags were rearranged and fluffed (the coach kept rolling twenty-four hours a day) until they neared the Platte River, and the flat prairies became cut with steep gullies.

"Every time we flew down one bank and scrambled up the other, our party got mixed somewhat," Sam recalled. "First we would all be down in a pile at the forward end of the stage, nearly in a sitting posture, and in a second we would shoot to the other end, and stand on our heads." Orion had considered it his civic duty to bring along a large, unabridged dictionary, which rode loose in the coach like a fourth passenger. "Every time we avalanched from one end of the stage to the other, the ... dictionary would come too, and every time it came it damaged somebody." Their pistols and coins settled to the bottom, but their "pipes, pipe stems, tobacco, and canteens clattered and floundered after the dictionary every time it made an assault on us, and aided and abetted the book by spilling tobacco in our eyes, and water down our backs. Still, all things considered, it was a very comfortable night."

The stage swung along the valley of the Platte, encountering a Pony Express rider near Scott's Bluff. "Away across the endless dead level of the prairie a black speck appears against the sky. ... In a second or two it becomes a horse and rider, rising and falling, rising and falling—

growing more and more distinct … nearer and still nearer, and the flutter of the hoofs comes faintly to the ear—another instant a whoop and a hurrah from our upper deck, a wave of the rider's hand, but no reply, and man and horse burst past our excited faces, and go winging away like a belated fragment of a storm."

Three months later the Pony Express was out of business, supplanted by a transcontinental telegraph line. It had carried the news of Lincoln's election in a remarkable six days from the Missouri to the Pacific. It bankrupted the company who founded it (and on whose stage young Sam Clemens rode west), but the memory of those glorious nineteen months of galloping ponies and fearless riders lived on long after the hoofbeats faded away.

The stage crossed the South Platte at Julesburg, Colorado, switching to a mud wagon and mules. The Concord coach was an oval eggshell engineered for lightweight strength, while mud wagons were heavy boxes on wheels, "a less sumptuous affair," Sam complained. River crossings were a hazardous undertaking, and the Platte was rising and treacherous with quicksand, but the mail had to go through. "Once or twice in midstream the wheels sunk into the yielding sands so threateningly … we half believed [that] we had dreaded and avoided the sea all our lives to be shipwrecked in a mud wagon in the middle of a desert at last. But we dragged through and sped away toward the setting sun."

They passed Fort Laramie in the night, the fur trading post on the North Platte which had become a government fort after the California rush increased traffic and potential conflicts with the reigning Sioux. "We had now reached hostile Indian country," Sam recounted, "and enjoyed great discomfort all the time we were in the neighborhood, being aware that many of the trees we dashed by at arm's length concealed a lurking Indian or two." But they passed on unmolested, crossing the North Platte and speeding toward the Sweetwater.

Sam noted the passing landmarks: Independence Rock, the oversized granite turtle shell with emigrant names crowding its polished surface; Devil's Gate where the river ran inexplicably through a mountain; Split Rock notched like a gun sight in the Granite Mountains; Ice Slough where the emigrants dug beneath the sod to find solid ice in July. They climbed Rocky Ridge, passed the soda lakes, clattered through Strawberry Creek and might have waved at me in my cabin had we inhabited a time similar to place.

On to the last crossing of the Sweetwater, where at South Pass Station, a "city" of four cabins, they were greeted by the hotelkeeper, postmaster, blacksmith, mayor, constable, city marshal, and principle citizen "all condensed into one person and crammed into one skin," what Bemis called "a perfect Allen's revolver of dignities." Said dignitary complained that "if he were to die as postmaster, or as blacksmith, or as postmaster and blacksmith both, the people might stand it, but if he were to die all over, it would be a frightful loss to the community."

Over South Pass to the Green River, looping down to Salt Lake City and across the Utah and Nevada deserts, "a thirsty, sweltering, longing, hateful reality!" Sam and the Secretary reached their destination of Carson City, capitol of Nevada Territory and close neighbor of the booming mining town of Virginia City—propped precariously on the side of a sagebrush hill—which Sam called "the 'livest' town, for its age and population, that America had ever produced."

The rush to Nevada had begun late in 1859, the result of a restless population and a leaked assay report of blue-gray quartz rich in silver and gold. The canyons around Mt. Davidson, in Nevada's elbow northeast of Lake Tahoe, had been worked since 1850 by a small group of misfit miners who placered their way uphill until they stumbled on the main lode, a two-and-a-half mile fissure filled unevenly with silver and gold that cut across the tall cone of Mount Davidson.

Two Irishmen, Pat McLaughlin and Peter O'Riley, made the first discovery, trying to separate gold out of the "blasted blue stuff" that clogged their rockers, unaware they were throwing away native silver. They had been working hard just to make wages, and suddenly their rockers were filled with a suspicious looking white gold. No one dreamed of silver in those days after California's golden bonanza. No one knew what it even looked like, a black and crumbly layer the Irishmen had struck while digging a reservoir to wash their pale gold.

H.T.P. Comstock had his suspicions, riding over the divide from his diggings on Gold Hill one day. Whatever the Irishmen had in their rockers looked good to him, and Comstock immediately informed them that they were working on his ground, and they'd better give him a share of their claim or get off. Henry Thomas Paige Comstock was a born promoter, especially of himself. His fellow miners called him "Old Pancake" because he was too busy to make bread, turning all his flour into pancakes. "And even as, with spoon in hand, he stirred up his pancake batter," according to Virginia City's chronicler, Dan De Quille, "it is said he kept one eye on the top of some distant peak and was lost in speculations in regard to the wealth in gold and silver that might rest somewhere beneath its rocky crest."

Tall and gaunt, his lower lip protruding with self-righteousness, Comstock was a big fish in the small pond of Johntown, the ramshackle collection of tents and huts in Gold Canyon. "Once Comstock got into the [Irishmen's] Ophir claim," De Quille reported, "he elected himself superintendent and was the man who did all of the heavy talking. He made himself so conspicuous on every occasion that he soon came to be considered not only the discoverer but almost the father of the lode." One other character of Johntown would descend to fame—"Old Virginia" Fennimore, who had scuttled to the remote side of the Sierras after a "difficulty" with a man on Kern River. Fennimore had a fondness for the bottle, and one night on a spree with the boys he fell and broke one, declaring "I christen this ground Virginia." And so the world-renowned silver vein became known as the Comstock Lode, and its chief metropolis, Virginia City.

O'Riley and McLaughlin continued to work their Ophir claim for gold, and it was only when a curious settler from the nearby Carson Valley sent a sample of the blue quartz across the Sierras to California for assay that the true value of the claim was learned. The results leaked out, a sworn secret passed from friend to intimate friend, and the stampede was on. One of the first over the mountains to buy a share in the Irishmen's Ophir Mine was George Hearst, whose good fortune would launch the great fortune of that newspaper family.

The Comstock Lode would ultimately produce three hundred million dollars in its twenty years of mining, and the fate of its discoverers offers an object lesson in human nature. To be a true prospector, you must always believe, somewhere in the back of your mind, that you could someday strike it rich. But what if the dream one day came true, and you let it all slip through your fingers? "Old

Virginia" Fennimore traded his claims on the Comstock for a horse, two blankets and a bottle of whiskey, and was later said to have complained that he owned a sixty-thousand-dollar horse but couldn't afford a saddle to ride it. He was riding it one day, on another spree, when he was thrown off and killed.

Patrick McLaughlin, who sold his share of the Ophir Mine for $3,500, turned up fifteen years later as a lowly cook at a California mine, and was ultimately buried at public expense. "Old Pancake" Comstock parted with his interest in the mine for $11,000. He went into the mercantile business but forgot to get his customers to pay for the merchandise, and eventually left town broke. In the 1860s he wrote a letter from Montana complaining that he had been cheated of his fortune. "I am a regular born mountaineer, and did not know the intrigues of civilized rascality," he lamented. But he thought the prospects looked pretty good in Montana, and believed he might once again strike it rich. At some point he changed his mind. On September 27, 1870, he shot himself in the head.

But perhaps the saddest story is of the second Irishman, Peter O'Riley. He thought he had something in the Ophir, and he held on longer than the others, finally selling for $40,000. "Had he placed his money at interest," De Quille commented, "he could have taken his ease all the rest of his days. But he built a big stone hotel in Virginia City and then allowed persons to persuade him that he was a great man, a man of financial genius, who should make himself felt in the stock-market. As he could neither read nor write, he was obliged to find persons to do that part of the business for him. He and his assistants then speculated—speculated until one day 'poor old Pete' found himself with pick, shovel and pan on his back, again going forth to prospect."

O'Riley put his faith in ghosts and spirits, who directed him to blast a tunnel into a bed of rotten granite in the Sierra foothills. "The ground in which he was at work was full of water, and cave-ins frequently occurred in his tunnel," De Quille reported. "The work of many weeks was often lost in a moment by a cave that crushed in his timbers and drove him back almost to where he first began; but the spirits said there was a whole mountain of silver and gold ahead, and he believed them and persevered.

"He was without money but not without friends. One and another of his friends among the old settlers purchased for him what he required in the way of provisions and tools. As he worked alone in the dark tunnel, month after month, far under the mountain, the spirits began to grow more and more familiar. They swarmed about him, advising him and directing the work. As he wielded pick and sledge, their voices came to him out of the darkness which walled in the light of his solitary candle, cheering him on; voices from the chinks in the rocks whispered to him stories of great masses of native silver at no great distance ahead, of caverns floored with silver and roofed with great arches hung with stalactites of pure silver and glittering native gold ... at last they boldly conversed with him under the broad light of day, and in the city as well as the solitude of the mountains. He was heard muttering to them as he walked the streets, and a wild and joyous light gleamed in his eyes as he listened to their promises. ...

"News at length came that O'Riley had been caved on and badly hurt," De Quille concludes, "then that the physicians had pronounced him insane ... he was sent to a private asylum ... and in a year or two died there." A man who believed he had something, he was someone, and he could be again.

The stampede to Nevada that followed the Irishmen's discovery in 1859 had already been rehearsed in California. Thousands had dashed south to Kern River in 1855, only to find the rumors of gold had been fabricated by speculators, an event that happened more than once in gold rush history. More than twenty thousand left in 1858 for the Fraser River in British Columbia, nine-tenths returning disappointed within a year. There had been gold there, but not enough to go around. Opportunities to get rich quick were gone in California, and with visions of '48 in their heads, the restless population was like a herd of nervous cattle in a thunderstorm, waiting for one golden bolt of lightning.

Only a few of the trampling stampeders made it across the Sierras before snow closed the passes to Nevada in 1859. Living conditions were primitive on Mount Davidson (originally named Sun Peak, then changed to honor an early California gold buyer). Lumber was hard to come by, and most houses were made of canvas or rough stone. "Many dug holes a few feet square in the sides of steep banks and, covering these with a roof of sagebrush and dirt, announced themselves 'at home' to their friends," De Quille reported. "As winter came on, not a few who had been living in tents or the open air, betook themselves for shelter to the tunnels they had begun to run into the hills, widening out a place at some distance back from the mouth for bedroom and parlor." All believed they would be rich men by spring.

What happened that spring when hordes began to pour over the Sierras is recorded by De Quille: "The few hardy first prospectors soon counted their neighbors by thousands and found eager and excited newcomers jostling them on every hand, planting stakes under their very noses and running lines round and through their brush shanties. ... The handful of old settlers found themselves strangers, almost in a single day, in their own land and their own dwellings."

J. Ross Browne was in the spring vanguard from California, and discovered on the slope of Mount Davidson "piles of goods and rubbish on craggy points, in the hollows, on the rocks, in the mud, in the snow, everywhere, scattered broadcast in pell-mell confusion, as if the clouds had suddenly burst overhead and rained down the dregs of all the flimsy, rickety, filthy little hovels and rubbish of merchandise that had ever undergone the process of evaporation from the earth since the days of Noah." It wasn't long before "the whole country was staked off to the distance of twenty or thirty miles. Every hillside was grubbed open, and even the desert was pegged, like the sole of a boot, with stakes designating claims."

The only thing missing from this mad rush was gold—the easy, available, everyman's opportunity scattered in streambeds just for the picking. The wealth of the Comstock was mostly in silver and wholly underground, which meant the common man had little chance of seeing any of it. Seventeen thousand claims were eventually recorded, but only a handful ever paid. The rush would have gone bust immediately were it not for the fabulous wealth of the few mines and the riding hope that a claim staked next to it or down the ridge would "come in" just as rich.

Without placer gold, the majority participated in the rush by trading in mining stocks. "Nobody had any money," Browne reported of early Virginia City, "yet everybody was a millionaire in silver claims. ...

All was silver underground, and deeds and mortgages on top; silver, silver everywhere, but scarce a dollar in coin." It was into this atmosphere of speculation that Sam Clemens came to Virginia City in 1862.

Since his arrival at Carson City the previous August, Sam had plunged with boyish enthusiasm into various speculative adventures. His first was a trek to Lake Tahoe with a pal to stake a 300-acre timber claim, requiring a fence around it and a house upon it to hold the claim. For a fence, Sam and his partner Johnny "cut down three trees apiece, and found it such heartbreaking work that we decided to 'rest our case' on those." They turned their attention to building "a substantial log house ... but by the time we had cut and trimmed the first log it seemed unnecessary to be so elaborate, and so we concluded to build it of saplings. However, two saplings, duly cut and trimmed, compelled recognition of the fact that a still modester architecture would satisfy the law, and so we concluded to build a 'brush' house.

"We devoted the next day to this work, but we did so much 'sitting around' and discussing that by the middle of the afternoon we had achieved only a halfway sort of affair which one of us had to watch while the other cut brush, lest if both turned our backs we might not be able to find it again, it had such a strong family resemblance to the surrounding vegetation." They camped on the lake shore, abandoning their house—"it was built to hold the ground, and that was enough. We did not wish to strain it"—and one day Sam's campfire got away from him, taking the timber claim with it. They watched the leaping wall of flames and decided to make their fortunes elsewhere.

Sam next joined three others in a trip two hundred miles north to the new Humboldt mines. They took two old horses and a wagon, and Sam might have been exaggerating when he said they had to take turns pushing the wagon *and* the horses. It only occurred to them later that if they had tied the horses behind, they would only have to push the wagon. At Humboldt, Sam learned the realities of prospecting, showing up at camp one day with a sparkling rock and believing his fortune finally made. "Granite rubbish and nasty glittering mica that isn't worth ten cents an acre!" his seasoned companion announced.

"So vanished my dream. So melted my wealth away. So toppled my airy castle to the earth and left me stricken and forlorn. Moralizing,

I observed, then, that 'all that glitters is not gold.' Mr. Ballou said I could go further than that, and lay it up among my treasure of knowledge, that *nothing* that glitters is gold." Sam left the Humboldt country no richer, but a little wiser as to the toil required to actually develop a mine.

In the spring of 1862 he was off to Esmeralda, a hundred miles southeast of the Comstock. With so much talk of instant wealth, Sam admitted that "I would have been more or less than human if I had not gone mad like the rest." He was certain that if brother Orion could send him more capital, they would eventually make a killing. "Twelve months, or twenty-four at furthest, will find all our earthly wishes satisfied, so far as money is concerned," Sam wrote in a letter. Orion forwarded what he could spare from his meager secretary's salary, but fortune continued to elude Sam.

In retrospect, Sam described those bitter and manic months at Esmeralda: "we were always hunting up new claims and doing a little work on them and then waiting for a buyer—who never came. We never found any ore that would yield more than fifty dollars a ton; and as the mills charged fifty dollars a ton for *working* ore and extracting the silver, our pocket money melted steadily away and none returned to take its place." Finally out of cash, he went to work in a quartz mill, shoveling, crushing, screening and separating ore.

"I will remark, in passing," Sam confided, "that I only remained in the milling business one week. I told my employer I could not stay longer without an advance in my wages. ... He said he was paying me ten dollars a week, and thought it a good round sum. How much did I want? I said about four hundred thousand dollars a month, and board, was about all I could reasonably ask, considering the hard times. I was ordered off the premises! And yet, when I look back to those days and call to mind the exceeding hardness of the labor I performed in that mill, I only regret that I did not ask him seven hundred thousand."

Sam had been a typesetter and sometime writer back in Mississippi, as well as a riverboat pilot, and a volunteer in the Confederate militia (for about a week), and it was only destitution that made him go back to writing. In August of 1862 he was offered a position on the Virginia City *Territorial Enterprise*, which he accepted with the misgivings of having failed as a miner. His job was "to go all over town and ask all sorts of people all sorts of questions, make notes of the information gained, and write them out for publication," admitting that he "let fancy get the upper hand of fact too often when there was a dearth of news." He massacred an entire imaginary family, made up hay wagons coming and going, and invented a petrified man sitting on a rock and thumbing his nose, a story widely reprinted as truth in other papers.

Virginia City in 1862 was a town floating on silver dreams, and nothing seemed truly impossible. Sam would learn his craft in the land of hyperbole, where suspension of disbelief was the order of the day, where humor, like everything else, was a matter of exaggeration. At times his elongated imagination got him into trouble; more often it brought him a kind of celebrity, and within six months he had chosen a pen name (like fellow reporter Dan De Quille, whose real name was William Wright) to suit his invention, and ever after was known as Mark Twain.

The twenty-one months Sam spent in Virginia City were flush times. Every wildcat mine (those off the main lode) printed stocks to sell and "every man owned 'feet' in fifty different wildcat mines and considered his fortune made." Sam, himself, owned a trunk half full of mine stocks, given to him for a favorable mention in the newspaper, or simply because it was the custom to give away stocks to friends. He would sell a few if he needed money, but most he hoarded away, waiting for the golden day they would be worth $1,000 a foot. In the meantime, his alter ego Mark Twain had sharpened his pen a little too pointedly and had been challenged to a duel. Sam decided to leave town on very short notice.

He landed in San Francisco and began living in the style which suited a potential rich man, keeping one eye on the stock boards and the other on fancy clothes and ritzy restaurants and popular nabobs. The market fluctuated wildly: "bankers, merchants, lawyers, doctors, mechanics, laborers, even the very washerwomen and servant girls, were putting up their earnings on silver stocks, and every sun that rose in the morning went down on paupers enriched and rich men beggared. What a gambling carnival it was! Gould & Curry [the most productive mine on the Comstock] soared to six thousand three hundred dollars a foot! And then—all of a sudden, out went the bottom and everything and everybody went to ruin and destruction! The wreck was complete. The bubble scarcely left a microscopic moisture behind. I was an early beggar and a thorough one. My hoarded stocks were not worth the paper they were printed on."

In borrasca, as the saying went, a dark and stormy time for Sam. "For two months my sole occupation was avoiding acquaintances; for during that time I did not earn a penny, or buy an article of any kind, or pay my board. I became very adept at 'slinking.' I slunk from back street to back street, I slunk away from approaching faces that looked familiar, I slunk to my meals ... and at midnight, after wanderings that were but slinkings away from cheerfulness and light, I slunk to my bed." Eventually Sam ran into an old mining friend and followed him to a once-booming camp in the California foothills, where only a few cabins remained of a "city" that had melted back into brush.

With the streams worked out, the few remaining dispirited miners were digging in the hillside paleoplacers for pockets of gold. They would wash a pan full of dirt, and if they found gold they would dig another pan to the side and then up, systematically tracing the washout back to the pocket. The pockets could hold thousands of dollars in gold, or nothing at all. "This is the most fascinating of all the different kinds of mining," Sam commented, "and furnishes a very handsome percentage of victims to the lunatic asylum." He pursued the fascination for three months, wandering from camp to camp and listening to the miners' stories. One was about a celebrated jumping frog. Sam took notes and rewrote the story and mailed it back East to the *Saturday Press* where it launched his career as a national writer, the only gold he would garner from California's hills.

When Sam left the West to pursue his fame as the garrulous Mark Twain, Virginia City was recovering from its stock market crash and a temporary failing in bonanza ore. A handful of mines turned out silver for another fifteen years on the Comstock, making a few people rich but paying break-even wages of four dollars a day to the several thousand miners working underground. The rush was over. For Sam it had been a downhill slide and a long climb back out to the white-suited gentleman with the drawling voice and bushy moustache who became so famous that he would stand one day, book in hand, in a shooting gallery in the near-ghost town of Virginia City, about as far from the Mississippi as you can get.

Chapter 7

I BELIEVED, as a child, in the literal truth of gold at the end of the rainbow. And I will tell you, if you don't already know it, that rainbows retreat when pursued, taking their gold with them.

The legend originated with the occasional discovery of tiny, dish-shaped coins in central Europe, made by the Celts before the rise of Rome. The Gnostic scholar Valentinus suggested they came from the place where the rainbow ends, mystical coins etched with simple stars and crosses, coins of the gods.

The Lydians, in Asia Minor, were the first mortals to strike coins, twenty-five hundred years ago. Their Pactolus River was rich in electrum, a natural gold-silver alloy, and it was a Lydian king who modeled for the phrase "as rich as Croesus." The Greeks told a story of an earlier Lydian king named Midas to explain the gold in the Pactolus River, how begging to be released from a wish that everything he touched turn to gold he should bathe in the river. The water took over his power and turned its own sands golden.

Greeks followed the Lydians in coin making, and Romans followed the Greeks. It turned out to be one of those "really good ideas" in history; over twenty thousand kinds of gold coins have been struck by more than twelve hundred cities and governments. Other items, from seashells to cigarettes, have been used as money to avoid a pure barter system of trade, but coins have been especially useful for several reasons. They were easily transported; their weight and purity could be standardized and their worth imprinted; and coins carried an intrinsic value in the content of their precious metal. They were not valuable simply because they could be exchanged for something else; they were worth, literally, their weight in gold.

Early coins were not uniformly round, but were made from lumps of equivalent size that were stamped with dies, and thus had the squashed appearance of a piece of thumb-flattened clay. It wasn't until the Renaissance that coins were struck from round blanks, with their edges milled (like our dimes and quarters) to prevent shaving. While the point of coinage was to standardize value, there were many ways, official and unofficial, to cheat. Citizens could shave a little off the edges, or "sweat" the coins by shaking a bag together and saving the dust to turn in for more coins. Rulers could subtly debase a coin, adding alloys or altering the weight. But such official depreciation was the hallmark of a failing regime. Near the end of the declining Roman Empire, for example, their silver coin was actually made of copper with a silver wash. Such coins were subject to "Gresham's Law," which simply states that bad money drives out good—people hoard good coins and pass along the bad at an increasing discount. When Gresham's Law ruled, it boded ill for the government and the governed.

Coins turned out to be useful for propaganda as well as trade, since anything could be stamped on them. The Greeks favored gods and goddesses, while the Romans preferred portraits of the emperor *au courant.* European monarchs also liked their faces on coins, but the true-to-life profiles were not always a pretty picture. Hungary's King Ferdinand had a markedly hooked nose, while Frenchman Louis XVI and Britain's Queen Victoria shared the hatchet profile of sharply receding forehead and chin. (After the French Revolution, that country notably beheaded their main coin and replaced the king with a tablet that read, "All men are equal before the law.")

The first American coins invariably portrayed Lady Liberty on one side and an eagle on the other, a democratic protest against autocratic rule. Colonial America had previously used Spanish "pieces of eight," which could literally be broken into eight bits to make change, (remember the ditty, "two bits, four bits, six bits, a dollar"?) which is why our quarter is sometimes called two bits. The word "dollar" is derived from a German coin called a *taler.* America broke away from its international monetary heritage by being the first country to issue coinage on the decimal system, a logical step that many nations have since followed.

America also issued some odd denominations that didn't survive, including a two-cent piece and a trime—a three-cent piece made of silver. A twenty-cent piece looked too much like a quarter, while a one-dollar gold coin issued during the California gold rush was smaller than our current dime and much too easy to lose. It was flattened and enlarged in 1856, but was dropped by the mint in 1890. Also issued in gold was a quarter eagle worth two and a half dollars, a half eagle at five, and an eagle worth ten dollars. The twenty-dollar double eagle was not minted until the California rush doubled the world monetary stock of gold. A hundred million double eagles were struck by the U.S. Mint in eighty-two years.

The coins that jingle in our pockets today are pale copies of their ancestors, redesigned since 1938, and robbed of their precious metal content in the 1960s. (Even the "copper" penny is zinc plated with bronze.) Against the Liberty tradition, Abraham Lincoln was the first real person ever depicted on a U.S. coin, supplanting a very Greek-looking Indian princess in 1909. The wheat ears on the back of that penny gave way to the Lincoln Memorial in 1959. Thomas Jefferson succeeded a buffalo on the nickel in 1938, and Washington bumped out Liberty on the quarter in 1932. Roosevelt replaced Liberty disguised as Mercury (wearing a winged cap) on the dime in 1946, Kennedy made the half dollar in 1964, within months of his assassination, and Eisenhower the dollar in 1971. Washington and his constitutional cohorts might be embarrassed that we have sunk so low, letting the government steal the real value of our money, and replace the symbolic ideal of liberty with their own unhappy heads.

The founding fathers of America intended coins to be the sole form of official money in this country. The Continental Congress had experimented with paper money to pay for the Revolutionary War, with disastrous results. Unbacked by metal, the notes became "not worth a Continental." Subsequently, all paper money was issued by private banks. The bank notes were sometimes backed by coin reserves and sometimes by thin air, leading to frequent failures. Citizens weren't required to accept bank notes as payment; whether they did depended on the bank's reputation and proximity.

But during the Civil War the government again resorted to printing paper money unredeemable in gold. The result was 250 percent inflation and a hoarding of coins that forced the government to print paper "coins" (called "shinplasters") in values less than a dollar so merchants could make change, an example of Gresham's Law at work. (As a side note, a bureaucrat in the treasury department had his portrait printed without authorization on the five-cent bill, prompting a law that prohibited living persons from appearing on American money. Before the law passed, the bureaucrat's boss managed to get his mug on the fifty-cent bill.) Paper "Demand Notes" printed during the Civil War used black ink on the front and green on the reverse, and so were called greenbacks.

By 1880 the U.S. government was able to back its paper money with gold, joining other countries in Europe to maintain a standard of exchange where each currency was valued at a fixed weight in gold. The Gold Standard at the turn of the nineteenth century maintained stable prices and forced a balance in trade between nations, but it also had a tendency to depress the economy and cause unemployment when there wasn't enough gold to go around. Congress had a raging debate, led by William Jennings Bryan, over the strangling effect of the Gold Standard, but new gold discoveries and improved milling techniques in the late 1890s helped to ease the gold crunch and quiet the critics.

The Gold Standard in Europe had to be suspended at the outbreak of World War I, but the U.S. held on until the Great Depression forced President Roosevelt to "nationalize" gold—all citizens were required to turn in their gold in exchange for paper. The dollar was then devalued by raising the official price of gold from $20 to $35 an ounce. American dollars were no longer redeemable in gold, but the currency was still backed by gold reserves. It is notable that American citizens held on to some $280 million in gold coins despite a law forbidding their possession, another example of Gresham's Law.

Repeated attempts to bring the industrial world back on a gold standard have largely failed, and in 1968 the U.S. dropped any requirement to back its dollar with gold. It is paper now and paper alone, backed only by faith in the government that issues it. What can happen to a currency backed only by faith in a government was well illustrated in Germany after World War I. In 1914 the German mark was equal to an American quarter; by November of 1923, with a defeated government unable to keep order, the mark reached four trillion to the dollar.

Imagine if one day you were a millionaire and the next your millions, trundled to the market in a wheelbarrow, would only buy a quart of milk. Germany sought refuge in a new, more powerful leader who promised to return that country to its former power and monetary stability, a leader whose name was Adolph Hitler.

Nowhere in the world is gold used as money now. It has become, in a sense, too valuable. People would hoard it, exchanging paper instead, and gold would flow out of government reserves and into private hands, a demonstration of our lack of faith, a hedge against disaster. William Jennings Bryan declared, in an 1896 protest against the monetary stringency of the Gold Standard, "you shall not crucify mankind upon a cross of gold!" And so we haven't. Something was lost in the bargain, though. The purchasing power of the dollar slips a little every year. You could, conceivably, save money your whole lifetime only to find it worthless in the end.

"Let everything go up in flames," wrote the poet Robert Browning, "what remains is gold." "A barbarous relic," proclaimed the economist John Maynard Keynes. And that it has become. We live now in a goldless society, forced to view the glowing metal through museum glass or on plated baubles in a jewelry store. Gold taken from the ground is returned to the ground, concentrated and bricked and hidden from view in guarded vaults. Gold is merely a metaphor today, an abstract of the thing itself, a word or color dashed down to spark the imagination, to evoke the qualities of richness and perfection.

I stood once in front of six gold bricks on display in the U.S. Mint in Denver, worth, that particular day, about a million dollars. I would like to see a whole wall of them glowing, as the Egyptians thought, like progeny of the sun.

APRIL 22. Winter has retreated here, hiding itself in the remnants of drifts, and everything looks softened and exposed. On south facing slopes buttercups are blooming unabashedly, with yellow petals shiny like enamel, the first sign of color, of spring.

The antelope are back now, following the receding snow. They have yet to fawn, knowing, like the tight-lipped willows, that there is risky weather to come. They run away from me half-heartedly, stopping to watch, then milling and jockeying like race horses in the first quarter mile. I saw a Swainson's hawk yesterday fighting the wind, two bluebirds and a killdeer the day before in the calm. Last night at dusk a beaver slapped her tail, a startling sound like someone throwing a big rock into the pond, "kawhoosh!" The moon rose an hour after dusk, glowing yellow through thin curtains of clouds that parted for starlight.

I can hear Strawberry Creek pouring through breaks in the beaver dams, artificial waterfalls that bubble and hiss. I have yet to find the beavers' lodge and must assume they live in a hole in the bank. During Pleistocene times beavers weighed 400 pounds or more, an unsettling thought. Now they average forty, but never stop growing. It is hard to gauge how big my beavers are because all I ever see of them is half a head and a streaming wake as they paddle around the pond. They will swim within a few feet of me if the wind is right, unable to see well.

In a way the beaver is a masterpiece of evolution, an overgrown mouse with webs on the hind feet but not on the front, self-sharpening teeth, transparent inner eyelids that act as diving goggles, a rudder for a tail. Their luxurious underfur, ideal for hat felt, nearly brought about their extinction in the past century, but they are alive and well now, at least in my pond.

I was admiring a half dozen ducks scooting around the pond, and a small bird in the willows I thought might be a female yellow warbler, when I heard a noise to the west, a sound like giant creaking doors bearing down from heaven. I looked up to see a pair of sand hill cranes winging along toward the Sweetwater and screeching like kamikaze pterodactyls sending radar signals to clear the runway.

So it seems the population is booming again here, up from moose, owl and magpie in March to hawk, cranes, ducks, bluebirds, antelope et al. A crow flew over today, loudly announcing her April arrival.

IT WAS APRIL when George Andrew Jackson finally revealed the secret he had kept through the frozen winter of 1859. The streams in the high Colorado mountains had begun to thaw, but were not yet raging with snow melt, and the south slopes were feeling sun after five months of moldering under drifts. The route was difficult—up the twisting canyon, through tangled timber, over rock slides and boulder heaps—but the goal was in sight. In his hand Jackson carried a crude treasure map of his own making. He found the small lodgepole pine with its top cut off, then the large fir tree marked with an axe blaze; he counted seventy-six steps due west, sank his shovel into a layer of ashes, and began to dig out gold.

Born on the Missouri frontier, George Jackson had already been to the California mines and back, had learned the skills of the mountain men and was friends with the old traders and Indians around Fort Laramie. He felt at home in the wild mountains, and was a good shot with his Hawken's rifle. He could tan hides and make his own buckskin clothing and moccasins, battle mountain lions and fight off wolverines. He could recognize a gold-bearing gravel bar when he saw one, and the six-foot-tall, seasoned veteran was only twenty-three years old.

Jackson first found his gold by accident, but it was the kind of fortuity that only happens to someone who is looking for two things at once. He had started up the tangled canyon of Clear Creek on New Year's Day, 1859, to hunt mountain sheep—his "supply of States grub short: two lbs. bread, half lb. coffee, half lb. salt," he reported in his diary. At Idaho Springs he found "1,000 sheep in sight tonight. No scarcity of meat in future for myself or dogs."

But the next morning his dogs woke him with low growls and the sheep were gone. "Mt. lion within twenty steps. Pulled my gun from under the blanket and shot too quick, broke his shoulder, but followed up and killed him." On January 3, the sheep came down again. "Are very tame; walk up to within one hundred yards of camp and stand and stamp at me and the dogs. Mt. lion killed one within three hundred yards of camp today, and scattered the whole band again."

Jackson explored the ascending canyon the next day, fighting deep snow. "Got back to camp after dark. Mountain lion stole all my meat today in camp; no supper tonight; D—m him." On January 5 Jackson was "up before day. Killed a fat sheep and wounded a Mt. lion before sunrise. Eat ribs for breakfast; drank last of my coffee. After breakfast moved up half mile to next creek on south side; made new camp under big fir tree. Good gravel here, looks like it carries gold. Wind had blown snow off the rim but gravel is hard frozen. Panned out two cups; no gold in either.

"Jan. 6. Pleasant day. Built big fire on rim rock to thaw the gravel; kept it up all day. Carcajou [a wolverine] came into camp while I was at fire. Dogs killed him after I broke his back with belt axe; H—l of a fight.

"Jan. 7. Clear day. Removed fire embers, and dug into rim on bed rock; panned out eight treaty cups of dirt, and found nothing but fine colors; ninth cup I got one nugget of coarse gold. Feel good tonight. Dogs don't. Drum is lame all over; sewed up gash in his leg tonight—Carcajou no good for dog.

"Jan. 8. … I've got the diggins at last, but can't be back in a week. Dogs can't travel. D—n a carcajou. Dug and panned today until my belt knife was worn out, so I will have to quit or use my skinning knife. I have about a half ounce of gold; so will quit and try to get back in the Spring."

Jackson marked the spot and drew a hasty map, then worked his way back down the rugged canyon along the iced-over creek bed, nineteen days out on what had begun as a week-long hunt for winter camp meat. He told no one but his partner Tom Golden (who founded the town of Golden, Colorado), "and as his mouth is tight as a No. 4 beaver trap, I am not uneasy." Jackson whiled away the winter, making a trip north to Fort Laramie to pick up mail for a dollar a letter. He played cards with his old trader friends who had congregated on the South Platte when rumors of gold had trickled like meltwater from the Colorado Rockies the previous summer. "Played poker all night with the boys," Jackson noted February 22. "I lost $20. Tom is mad." Four days later Tom was still mad, and on the 27th the two parted company.

When Jackson finally returned to his buried treasure in April, he took with him a group from Chicago who had "the best supply of grub and mining tools of any company in the country," and left Tom Golden to look for his own fortune. Jackson and the Chicago boys walked into the new town of Denver in May, and laid down $1,900 in gold dust from a week's worth of mining. The news of Jackson's discovery, and two others in the spring of 1859, came just in time to save the "Pike's Peak Rush" from total ruin.

The idea of gold in Colorado was not a new one. Spaniards had prospected as far north as Long's Peak in the eighteenth century, and trappers in the 1830s had sometimes brought gold as well as furs into the trading posts at Santa Fe, Laramie, and Bent's Fort. By the 1840s it seemed to be common knowledge that there was some gold in the headwaters of the South Platte and Arkansas rivers. In 1850 a party of Cherokees and whites with mining experience from the gold fields of Georgia were on their way to California by what would become the Cherokee Trail (through Kansas, Colorado and southern Wyoming) when they stopped briefly at a tributary of the South Platte to wash fine gold from the sands.

The Cherokees would remember that gold long after they tried their luck in California and returned home again. They talked about it for eight years, told their friends and neighbors who dreamed of sudden wealth, of easy riches. Seventy of the dreamers set out for Colorado in the spring of 1858, and landed that June on Cherry Creek where it flowed into the South Platte. They expected, wrote member Luke Tierney, "to find lumps of gold like hail-stones, all over the surface. . . . The prospects fell so far short of their sanguine expectations and feverish hopes, that many began to show evident signs of disappointment and mortification." By July all but thirteen had gone home again in disgust.

William Green Russell, a veteran of both the Georgia and California gold fields, refused to give up so easily and became the leader of the tenacious little group who stayed behind. They prospected along the Platte upstream from Cherry Creek, and found a few good pockets of placer gold in late July and August. Word of their small success filtered up to Fort Laramie, where George Jackson heard about it and started south, as did a whiskey trader who visited the Colorado camp and carried the word east, along with a small sample of fine gold and a fairly rich bag of unwashed gravel.

"The New Eldorado!!!" shouted the headline of the Kansas City *Journal of Commerce* on August 26, 1858. A few ounces of gold was turned into pounds, and reminiscent of California in '48, the *Journal* reported that "men have dug out over $600 per week, with nothing but their knives, tomahawks and frying pans. . . . We have no hesitation in saying that in a fortnight the road from Kansas City . . . west to the mines will be lined with gold seekers." But the public vacillated, wavering on that thin line between credulity and disbelief. Newspapers like the *Kansas Weekly Press* were circulating stories of a man who had returned with a whole kettle full of gold worth six to seven thousand dollars, or a small boy who dug out a thousand dollars and could get "all he wants." It was good for business, and it was good, they all thought, for America.

But "the plain, simple truth of the matter is this," Luke Tierney warned, "an industrious man can make from two dollars and fifty cents to fifteen dollars a day. ... Richer diggings may, of course, be discovered hereafter, where fifty dollars a day can be realized; but no such places are yet found." Tierney and the rest of the Cherokee party quickly exhausted the pockets they had found, and set out in an unsuccessful search for more. When Green Russell returned to Georgia for reinforcements that October, he took with him the party's total summer earnings, a mere $500 worth of fine dust.

But where gold is concerned, good news travels faster and farther than bad. Word of the new gold strike had already spread up and down the Missouri and back to the major eastern cities, and small parties were leaving daily for the new mines. By December a thousand gold seekers reached Cherry Creek, where speculators had already laid out the twin towns of Auraria (after Green Russell's hometown in Georgia) and Denver (named for the governor of Kansas) on opposite sides of the creek.

Residents with an optimistic nature or a speculative interest in the new town sites sent back letters that fueled the fire. Yes there was gold everywhere, they reported. While they, personally, hadn't got their hands on any of it, they heard that others were making from two to ten dollars a day. "There is gold, and mining will pay well to those who are able and willing to work, and to such I say by all means come," penned one correspondent in December. But the decamped mayor of Nebraska City disagreed. "We were quite surprised a few days since," he wrote from Denver in January of 1859, "when we read the glowing accounts in the Missouri River papers, of what the miners are doing out here. I pronounce them a pack of lies, written and reported back by a set of petty, one-horse-town speculators, and are calculated to ruin many a poor devil besides your humble servant."

Cold weather had shut down most attempts at mining that winter, so the earliest stampeders had to wait and see. "There is no doubt about the gold," one wrote back cautiously, "but the question is whether it will pay." It was a question those on the Missouri frontier didn't dwell on. While newspapers printed a few accurate stories of the limited extent of gold found in what was then Kansas Territory, they were outweighed and outnumbered by exaggeration and hyperbole. In their minds the Fifty-niners would be just like the Forty-niners, rolling in gold.

The rush that followed was almost inevitable, for all the necessary ingredients were assembled in the spring of 1859: a two-year depression had brought hardship and a labor surplus to the Midwest; there was a lack of competing attractions elsewhere (the rush to Nevada drew participants mainly from California); and only six hundred miles away lay a spectacularly publicized (though not spectacular) gold discovery.

But the Fifty-niners would differ from the Forty-niners in several significant ways. Those bound for California could not easily turn back and rarely did. The existence of a tremendous amount of gold was proven before the Forty-niners set out, and because the long trip required a fair amount of capital to undertake, most of the California argonauts were not the poorest and most in-need citizens of the country. But the Fifty-niners were often both poor and desperate. Two hundred dollars would buy a good outfit of wagon, oxen and supplies for the forty-day crossing of six hundred miles of prairie between the Missouri River and Cherry Creek. Many could afford nothing more than resoled boots and a bag of flour. "Hundreds are now starting on foot," a dispatch from St. Joseph in March of 1859 reported, "with nothing but a cotton sack and a few pounds of crackers and meat, and many with hand-carts and wheelbarrows. ... They are the poorest of creation."

Invention supplanted means for many of the stampeders. In February two Minnesota men passed through Council Bluffs pulling a toboggan to the mines. An Illinois man set out with a dog team. "He has a light wagon, two Newfoundland dogs, two greyhounds and two pointers for the lead, and expects to distance all competition," the papers reported. A man in Westport designed a widely talked-about wind wagon, complete with sails, to carry passengers across the plains in an advertised six days.

In April, "a full rigged band of musicians from Indianapolis" passed through Kansas City. "What a commotion the saxhorns, trombones, kent-bugle and bass drum, will raise among the Kiowas and Cheyennes in the valleys of the Rocky Mountains" the reporter thought. By the end of April the *Westport Border Star* reported that "crowds continue to come—all sorts, sizes and descriptions. The world seems all a moving." Among the movers, according to said paper, were a short-cut seeker from Illinois wanting to get to Colorado in a week, a Green Mountain boy with a stump-tailed mule and a crumpled-horn cow, then a "rip-snorter" from Tennessee, and six St. Louis "rounders and runners."

They were followed by "a tidy little red wagon, drawn by four sleek, well-fed oxen, and followed by a little muley cow, with a merry bell hanging from her neck, of which she seems to be proud, for every now and then she takes occasion to give her head an extra shake just for the music of the thing. The wagon body is compactly stowed with a judicious lot of emigrants' outfitting articles, and in the front, on a calico-cushioned stool, backed up by a pile of clothes and bedding, sits a buxom, blooming young woman, holding up a bright and crowing baby … to the admiring view of the stout-limbed and sun-browned young man walking abreast the wheel-oxen who seems to divide his attention between his team and his load, and cracks his long whip as much for the delight of his baby as for the encouragement of his cattle."

In contrast to the pretty red wagon, the *Missouri Republican* presented this description of a party en route to the mines. "The man was driving an overloaded wagon, while the woman was hauling the sick child in a hand-cart, carrying the medicine in her hands. In their hurry to reach the land of gold, they could not stop long enough for their invalid offspring to recover its shattered health."

All were bound for the "Pike's Peak Mines." Even though Pike's Peak was eighty miles from Cherry Creek and goldless itself, it was the only landmark at that far edge of Kansas Territory that most people had heard of. "Pike's Peak is in everybody's mouth and thoughts," the *Missouri Republican* reported, "and Pike's Peak figures in a million dreams. Every clothing store is a depot for outfits for Pike's Peak. There are Pike's Peak hats, and Pike's Peak guns, Pike's Peak boots, Pike's Peak shovels, and Pike's Peak goodness-knows-what-all, designed expressly for the use of emigrants and miners, and earnestly recommended to those contemplating a journey to the gold regions of Pike's Peak. We presume there are, or will be, Pike's Peak pills, manufactured with exclusive reference to the diseases of Cherry valley, and sold in conjunction with Pike's Peak guide books; or Pike's Peak Schnapps to give tone to the stomachs of over tasked gold diggers; or Pike's Peak goggles to keep the gold dust out of the eyes of the fortune hunters."

Outfitting towns all along the Missouri were vying for Pike's Peak business, each claiming to have the cheapest supplies and the best trails out. Some of the advertised trails were familiar and well documented. The Santa Fe Trail from St. Louis followed the Arkansas River, and by turning north instead of south, led to the headwaters of Cherry Creek. The Oregon Trail following the Platte from Omaha had become as worn and familiar as an old shoe. By turning along the South Platte instead of the North, wagons rumbled right into the Cherry Creek junction and the new town of Denver.

Between the northern Oregon Trail and the southern Santa Fe road was a no-man's land of sand hills and tendril-like drainages. In the spring of 1859 the Pike's Peak Express Company pieced together a stage line south of the Republican River, with twenty-six way stations between Leavenworth and Denver. But many Fifty-niners were too impatient to wait for the beginning of stage service, or too poor to afford the fares. The shortest, most direct advertised route to Pike's Peak was along the Smoky Hill River, and thousands set out from Leavenworth or Kansas City expecting a quick, easy trip, some with ox teams and lightly loaded wagons, some with a single pack horse, and some walking with a pack on their backs. But the Smoky Hill "trail" was a figment of its promoters' imaginations, a line drawn easily on the map, much like Hastings Cut-off had been drawn across Utah for the Donner party.

Letters trickling back to Kansas newspapers in May related the difficulties. "We left Leavenworth full of confidence in what is known as the Smoky Hill route," wrote one correspondent. "On reaching the Saline—two hundred and sixty miles—we lost all track of a route, and were forced to make our own." There was too little grass on the route to keep oxen alive, and too little food packed because of the advertised ease.

J. Heywood, who took a stagecoach from Leavenworth, reported a great deal of suffering among those who had chosen the Smoky Hill road. "They have got lost, run out of provisions, and actually starved to death. On Wednesday evening, the fourth of May, we camped on Beaver Creek, some sixty miles east of Denver. While there a man was brought into camp by an Arapaho Indian almost dead for want of food. He had eaten up two of his brothers."

Alexander, Charles, and Daniel Blue had left Whiteside County, Illinois, in February, holding the highest of golden hopes. They had read all the stories of depthless treasure in the Pike's Peak country, and if the press had learned nothing else from the California experience, it had learned how to ignite people's dreams. The Blue brothers picked the shortest trail to Eldorado, leaving Kansas City with a single pack horse of supplies. Twenty days out they lost their horse, and then their way. Alexander, the oldest brother, was the first to die. "At his own last request, we used a portion of his body as food on the spot, and with the balance resumed our journey," Daniel Blue confessed. "We succeeded in traveling but ten miles, when my youngest brother, Charles, gave out. ... I also consumed the greater portion of his remains."

Heywood described the rescued survivor after he was brought into camp by the Arapaho. "The poor sufferer was almost a skeleton. His cheeks were sunken and his eyes protruded from their sockets. He looked wild and had almost lost his sight. At times his mind would wander, and he seemed to be very much affected." Blue was loaded onto the stagecoach and carried for the last leg of his once hopeful journey into Denver.

How especially disappointed must the Smoky Hill travelers have been to find the streets of Denver paved not in gold, but in mud. One young man, after suffering terribly on the route, wrote home to his father that "there is no gold at Pike's Peak. No man can make ten cents a month. I am out of money, and without a chance to make any. Therefore, dear father, send me one hundred and twenty-five dollars to take me back home."

The gold that had been found on the South Platte, Cherry Creek, and all the lower streams was fine as flour and unevenly located. It turned out there was not enough for the first thousand arrivals, let alone the hundred thousand who were on their way west. "Many persons will rake and scrape up all the money they can gather, and proceed to the gold regions," warned a prophetic editorial back in September, "where they will probably meet only with disappointment, spend their means and be left destitute." The poorest of the rush, the handcart haulers and pack pony trains, wheel barrow pushers and foot-worn hikers arrived at Cherry Creek first, and found a rag-tag town short on supplies, shorter on employment, and shortest on gold. They begged what food they could and started home, many trying to float down the Platte where their homemade boats capsized or ran aground on the sand bars. They told everyone they met that Pike's Peak was a humbug, and the rush began turning back on itself in a whirlpool of malediction and retreat.

Of the hundred thousand who set out for Colorado, half turned back before reaching Denver, and another quarter returned after they arrived at Cherry Creek and confirmed the bad news for themselves. At Fort Kearny on the Oregon Trail, E.H.N. Patterson encountered a "wild panic" the beginning of May. "Five teams are going back where one is going forward," he wrote of the confused multitude. Many were blaming guidebook author D.C. Oakes for encouraging their folly in joining the rush. He was hung in effigy, and buried on the plains with a tombstone reading "Here lie the remains of D.C. Oakes, who was the starter of this damned hoax!" Patterson reported passing such a grave, and nearby, "alive and ... kicking," the villain, Oakes himself, on the way back to the mines he had left in the fall to write his guidebook of promises. But Patterson, a thirty-year-old newspaper editor and California gold rush veteran from Illinois, was not easily dissuaded, and so pushed on against the ebbing tide, determined to see for himself.

Charles Post was in the middle of Kansas, on the Santa Fe Trail, when he met "a large train going back, having heard bad news from Pike's Peak." But, he added, "it is a fact that no one we have yet met have struck a shovel in the ground for gold, but have taken others' say so for it," and so the twenty-seven-year-old lawyer from Michigan also pushed on to see for himself. Post had encountered the infamous Wind Wagon abandoned outside Kansas City. "It is a four-wheeled vehicle about nine feet across, schooner rigged [with] a very large sail. The whole weighs three thousand pounds. It plowed right through the mud, but cast anchor in a deep ravine where the wind failed to fill the sail and she stopped."

On the stage route of the Pike's Peak Express, A.D. Richardson noted "many returning emigrants, who declare the mines a humbug; but [we] pass hundreds of undismayed gold seekers still pressing on." One stranded stampeder had crossed out "Pike's Peak or Bust" on his wagon, and charcoaled in "Busted, by thunder!"

At one of the stage stops, Richardson noticed "a fair, delicate Indiana boy who ran away last spring, froze his feet en route for the mines, and after many hardships is now glad to return to home and school." But the boy turned out to be a girl. "She was dressed in male costume with a slouching hat which she wore at table to conceal her features. She talked little but in walking from the tent to the coach her gait betrayed her. She is twenty years old; appears intelligent and well educated; professes to be returning to her parents in Indiana after spending three months in the mines; but gives no reason for her dangerous and unwomanly freak."

By May the *New York Times* was printing the bad news. "We hear nothing of gold, but much disappointment, suffering and repentance. Miners are leaving in crowds. Pike's Peak is no California. ... They will return, some of them as can, stripped of everything, to begin again the labors from which visions of gold allured them." The rush had hit rock bottom, and amid the swirling confusion of go-backs and go-forwards, George Jackson walked into town with a bag full of gold, and so did John Gregory.

Gregory was a Georgian miner who had been on his way to the short-lived Fraser River rush in B.C. when circumstances forced him to winter in Fort Laramie. Like Jackson, Gregory had followed up on the gilded rumors from Cherry Creek, prospecting all the lower streams to see which was most promising. He worked his way up the north fork of Clear Creek and found a gulch that yielded four dollars in the first pan. Gregory washed out nearly $1,000 in five days, then sold his claims for $21,000 and his talent at prospecting for $200 a day.

Within a month five thousand miners had scrambled up to the Gregory Diggings where the towns of Central City, Blackhawk and Nevadaville formed one long string of sudden civilization. The rock walls of the canyon were splintered with hundreds of quartz veins, many containing gold, which had weathered to a crumbly mass of easily worked gravel. Fifty dollars a day per man, the golden promise that had lured a hundred thousand from their two-dollar-a-day-or-less occupations, had finally been realized.

One of the lucky strikers was Green Russell, that "old Californian" from Georgia who had persevered in 1858 when most of his Cherokee party went home. He knew that the "float" gold on the Platte had to come from higher in the mountains. The finer the gold, as a rule, the farther it has traveled, ground down in the mill of churning stones and rushing water. (Additionally, the finer the gold the farther it *can* travel, carried in suspension for hundreds of miles to be left on a sandbar in the most unlikely of places.) When Russell tried to pick his way up Clear Creek that first summer, the terrain proved too difficult. "A bird couldn't fly up that canyon," he declared. Once Gregory showed the way, Russell discovered his own rich gulch, which soon had nine hundred men taking out $35,000 a week.

By late June, Denver was largely deserted, much as San Francisco had been at the start of the California rush; once-valuable town lots couldn't be given away. "The arrivals from the states still continue to be large, but returning emigrants have become a scarce article," wrote one Denver correspondent. "Whoever lands here, now at once steers for the mountains." E.H.N. Patterson, who had traveled on to Colorado despite the "panic" of returning Fifty-niners, passed through the quiet streets of Denver to find strung along Clear Creek "thousands of cattle, hundreds of wagons and tents, and people innumerable [thronging] the valley of the stream."

Patterson described the harrowing pack trail up to Central City. "It leads along the sides of the high mountains that fall off with a slope of seventy-five degrees—traverses deep valleys—runs over lofty hills—winds about through pine woods and among rocky ledges." The road was "now steep, now sideling, now running over boulders and steep ledges of rock, and leading through deep mud holes, and winding about among dense groves of aspen and pine." At the end of the road was a deluge of men picking and shoveling and rushing pell-mell with stakes and pans.

The flood loosed by these first discoveries soon spread to other drainages and washed up rich strikes along the way. The upper tributaries of Boulder Creek to the north had been worked since early spring, and on the South Fork of Clear Creek, upstream from the Jackson Diggings at Idaho Springs, new finds were made at Empire and Georgetown. Rich placers were uncovered in South Park in July. On Tarryall Creek the discoverers (who could make up their own mining laws), gave themselves 150 feet of stream each rather than the traditional 100 feet. This so disgusted latecomers that they referred to it as "Grab All Creek" and started their own town across the valley, called Fairplay.

By August prospectors had crossed the divide to the Blue River drainage where they found "pound diggings"—a pound of gold per man per day—near Breckenridge. That fall, others worked their way up the Arkansas River where the following spring California Gulch would pay spectacularly, like the days of '48, all along its seven-mile course. The better claims yielded $65,000 that season. A two-million-dollar production in 1860 doubled the output of 1859, and peaked in 1862 at $3.5 million with a decade total of $25 million.

But the land of found gold was the land of lost souls as far as Reverend John Dyer was concerned. The "snowshoe itinerant" had walked to Colorado from Minnesota in 1861, to save the miners from the wickedness of their circumstances. It wasn't gold digging that was evil (for Dyer was not above digging himself), it was the way they spent their golden profits in drinking and dancing. Theaters were third on his list. Reverend Dyer rode the circuit (or more often walked since he could rarely afford a horse) from South Park to Breckenridge to California Gulch and back as a Methodist minister. "In about four months I traveled near five hundred miles on foot, by Indian trails, crossing logs, carrying my pack, and preaching about three times a week. Received forty-three dollars in collections at different places. … Spent about fifty dollars of my own resources, as I had worked by the day and job through the week, and preached nights and Sundays. My clothes were worn out; my hat-rim patched with dressed antelope skin; my boots half-soled with rawhide," Dyer summarized. When he left Minnesota Dyer weighed 192 pounds. After six months in the mines he weighed 163. "I found out that a man at forty-seven, getting fat, could walk, work, and preach off all the fat," he announced.

It wasn't always easy to find people to preach to. Having hiked eight miles from Buckskin Joe to Montgomery, "instead of finding a good number, all were out staking off claims but one man." It was easier to find places to preach in—there were no churches in the new communities so Reverend Dyer would simply pick a shady tree and arrange several logs for benches. Services were frequently held around a simple campfire, or in someone's house. At Washington Gulch near Breckenridge, Dyer commenced the service in a tent that was also a grocery and saloon. In the middle of his sermon "a mule reached his head into the tent and took out a loaf of bread and started off with it." The mule was caught and forced to give up his prize, and the service continued.

The congregation was plentiful in Breckenridge because Dyer made sure of it. "The way we got them out was to go along the gulches and tell the people in their cabins and saloons where the preaching would be at night, and then, just before the time, to step to the door where they were at cards, and say: 'My friends, can't you close your game in ten minutes, and come and hear preaching?'"

At Buckskin Joe, where he had a cabin, Dyer also reported success despite "every kind of opposition—at least two balls a week, a dancing school, a one-horse theater, [and] two men shot." But it was difficult to support himself as an itinerant preacher. Collections were niggardly when taken at all, the church offered little missionary support, and a man had to have a strong calling to pursue such work to his own impoverishment. When Dyer's mining efforts failed, he took a contract to deliver mail from South Park across thirteen-thousand-foot Mosquito Pass to California Gulch once a week all winter long. His snowshoes were long, Norwegian-style board skis, and his brushes with cold death were frequent since he had to cross the pass at night while the snow was frozen hard enough to support him. Dyer carried thirty-five pounds of mail each trip, and also offered his services as express agent, exchanging discounted greenbacks for the gold that flowed freely from California Gulch. It was the most money he ever had in his life, twelve hundred dollars.

Father Joseph Machebeuf of the Catholic Church also had trouble squeezing money from the miners. When he arrived in Denver in the fall of 1860, his first job was to finish the church building at Fifteenth and Stout streets, an empty prairie then, and downtown Denver now. It was a thirty-by-forty-six-foot, half-walled brick structure, and it took him two months to finally get the roof on. "It was not plastered and was without windows, but the mountain evergreens hid the rough walls ... and canvas kept out some of the cold wind," Machebeuf's biographer described. Father Joseph had been sent up from Santa Fe when news of the gold rush reached there, and when he wanted money or articles for his Colorado churches he would return to the poor and pious New Mexicans who were more generous than his own gold-maddened flock.

Machebeuf, who was the model for Willa Cather's Father Joseph Vaillant in *Death Comes for the Archbishop*, lived in a shack at the rear of his church, and spent most of his time, like Reverend Dyer, traveling from one mining camp to the next. Unlike Reverend Dyer, Father Joseph traveled by buggy, his own primitive mobile home that carried "his vestments for mass, his bedding, grain for his horses, his own provisions and his frying pan and coffee pot." He had brought the buggy from Santa Fe, and it was "of a peculiar shape, with square top, side curtains, a half curtain in front to be let down in cases of storms, and a rack behind for heavy luggage. It was not long before it was known in every camp, and the sight of it was sufficient notice to the people that the priest had come."

Father Joseph found only ten Catholic families living in Denver when he arrived, with some two hundred more transients. The mobility and restless nature of gold rushers made church building a hard job, but at Central City Father Joseph succeeded because "one Sunday at the close of the mass he had the doors locked and the keys brought to him at the altar. Then he declared that no one would be permitted to leave the hall until the question of a church was settled."

Father Joseph continued his attempts to build Catholic churches in settlements that would become permanent, but by 1864 most of the great Colorado gold camps were ghost towns. The placer gold had been rich in places, but was limited in extent and quickly worked out.

The miners moved on to day wages in the hard rock mines of Central City, or they floated away on rumors of new diggings in Idaho and Montana. Reverend Dyer reported the exodus from his district at the end of 1863 and remarked that "what members remained were so poor that they could not get away."

"The relics of former life and business," Ovando Hollister reported from abandoned California Gulch, "old boots and clothes, cooking utensils, rude house furniture, tin cans, gold pans, worn-out shovels and picks and the remains of toms, half buried sluices and riffle boxes, dirt roofed log cabins tumbling down and the country turned inside out and disguised with rubbish of every description, are most disagreeably abundant and suggestive."

Even around Central City, where John Gregory had first washed out a four-dollar panful and Green Russell had finally made his pile, troubled times were threatening. The rich lodes, oxidized and crumbled enough at the surface to be worked with only a sluice box in the spring of 1859, turned to hard rock by fall, requiring blasting to remove the ore, and steam-powered stamp mills to crush it. At a depth of only thirty feet the ore's true character revealed itself as a mixture of gold, silver and sulfides which refused to yield to traditional milling methods using mercury. By 1864, seventy-five percent of the gold was washing out with the tailings and nobody knew what to do about it.

In 1866, Bayard Taylor—the travel writer who had chronicled California's days of '49, (meeting the infamous "Buckshot" and resisting the glittering urge to dig gold)—took a summer trip to Colorado to see what he could see. At Central City he found that "the deserted mills, the idle wheels, and the empty shafts and drifts for miles along this and the adjoining ravines—the general decrease of population everywhere in the mountains—indicate a period of doubt and transition, which is now, I believe, on the point of passing away." But it would take another decade and a different metal—silver—to revive the Colorado camps. "Colorado has been, alternately, the scene of extravagant hopes and equally extravagant disappointments," he concluded.

The wreckage left behind was more persistent. "The view of the intersecting ravines ... and the steep, ponderous mountains which inclose them, has a certain largeness and breadth of effect, but is by no means picturesque," Taylor noted. "The timber has been wholly cut away, except upon some of the more distant steeps, where its dark green is streaked with ghastly marks of fire. The great, awkwardly rounded mountains are cut up and down by the lines of paying lodes, and pitted all over by the holes and heaps of rocks made either by prospectors or to secure claims. Nature seems to be suffering from an attack of confluent small-pox. My experience in California taught me that gold mining utterly ruins the appearance of a country, and therefore I am not surprised at what I see here."

Leaving Central City, Taylor took the stage across the divide to the south fork of Clear Creek, passing Russell Gulch that had paid its namesake so handsomely. The gulch "from top to bottom, a distance, apparently of two or three miles, and all its branches, show the traces of gold washing. The soil has been turned upside down, hollowed out and burrowed into in every direction."

Idaho City, at George Jackson's discovery, was a "straggling village of log huts," its hot springs already turned into bath houses for summer tourists. The height of the mountains towering above the valley deeply impressed Taylor. "Take the altitude of the Catskill Mountain House above the Hudson, and place that on top of Mount Washington, and you will have the elevation of this place, where people live, work, and carry on business."

Much of Colorado was still wild and forbidding, Taylor discovered as he crossed Berthoud Pass into Middle Park. "We found ourselves at the head of a superb meadow stretching westward for five or six miles, bounded on the north, first by low gray hills of fantastic shape, then by great green ascending slopes of forest, and above all, jagged ranges of rock and snow."

Circling around Middle Park, Taylor followed the Blue River up to Breckenridge. "Over ditches, heaps of stone and gravel, and all the usual debris of gulch mining, we rode toward some cabins which beckoned to us through scattered clumps of pine. A flag-staff, with something white at half-mast; canvas-covered wagons in the shade; a long stretch of log houses; signs of 'Boarding,' 'Miner's Home,' and 'Saloon,' and a motley group of rough individuals ... such was Breckenridge!"

Taylor strolled the rugged streets, discovering the white "truce flag" was actually a faded and tattered American flag in front of a log court house and lowered to half mast in "comical mourning" because "the bully of Breckenridge—a German grocer—had been whipped, the day before, by the bully of Buffalo Flats."

Only five hundred miners remained of the five thousand who had flocked to Blue River six years earlier in search of "pound diggings." (The Weaver brothers, discoverers of nearby Gold Run, took away ninety-six pounds of gold after one season of mining.) Breckenridge had originally been spelled Breckinridge, after James Buchanan's vice president, but when John Breckinridge lost the 1860 presidential election to Lincoln, then joined the Confederate cause, Unionists in the mining camp changed the first *i* to an *e* to disassociate themselves from the traitor.

Bayard Taylor left Breckenridge to cross 11,541-foot Hoosier Pass, which dropped down to the headwaters of the South Platte and the broad and beautiful South Park where Tarryall and Fairplay had carried on their rivalry. Taylor stopped to visit Montgomery at the foot of the pass. "It once had a population of three thousand, and now numbers three or four hundred. But as the cabins of those who left speedily became the firewood of those who remained, there are no apparent signs of decay."

Up a side valley, Taylor rode on to Buckskin Joe, named for a leather-wearing miner named Joe Higginbotham who found placer gold there in 1860. The town boomed while the Phillips Lode turned out half a million dollars from a quartz vein. "The people, for the space of two or three years, made a desperate attempt to change the name to 'Laurette,'" Taylor commented, "but failed completely, and it will probably be Buckskin Joe to the end of time. ... There are worse names in California than this, and worse places."

Taylor next crossed Reverend Dyer's nemesis, Mosquito Pass, to the Arkansas River where Leadville would anchor the silver boom that began in 1878. That "blasted blue stuff" would clog the sluices of the first miners in the gulches, a carbonate of lead and silver that was thrown away as a nuisance in the mad scramble for gold. Abe Lee, the discoverer of the best-paying gulch in the area, was prospecting under snow when he yelled "Boys, I've got all California here in this pan!"

California Gulch was still paying a hundred dollars a day per man on some claims when Taylor visited the metropolis of Oro City—like most of the other camps, a collection of log cabins, "then a street with several saloons, eating houses, and corrals." The town was celebrating Independence Day when Taylor arrived July 4 and delivered a lecture in the recorder's office, which was draped in flags and turned into an auditorium. "After the lecture there was a ball," Taylor described, "which all the ladies of the Upper Arkansas Valley—hardly a baker's dozen—attended."

Leaving California Gulch, Taylor followed the Arkansas downriver through its colorful canyon. "The succession of tints is enchanting as the eye travels upward from the wonderful sage-gray of the Arkansas bottom, over the misty sea-gray of the slopes of buffalo grass, the dark purplish green of fir forests, the red of rocky walls, scored with thousand-fold lines of shadow, and rests at last on snows that dazzle with their cool whiteness on the opposite peaks, but stretch into rosy dimness far to the south."

Crossing back through South Park, Taylor returned to Denver and finished his looping tour of the Colorado mines—all enchanting scenery and in between the broken-down leftovers of a hundred thousand dreams. Despite the mining bust, Denver was still building with the blind optimism it had shown from the beginning. Taylor estimated the population at six thousand in 1866. "Probably no town in the country ever grew up under such discouraging circumstances, or has made more solid progress in the same length of time. It was once swept away by the inundation of Cherry Creek; once or twice burned; threatened with Secession; [and] cut off from intercourse with the East by Indian outbreaks," Taylor concluded.

It was 1864 that three-year-old Colorado Territory began having serious Indian problems. Relations between the Indians and whites had been relatively stable since the Pike's Peak rush began, a tribute to restraint on both sides. The miners had invaded land officially ceded to the Cheyenne and Arapaho by the Fort Laramie Treaty of 1851, a huge tract stretching from Kansas to the Rockies between the Platte and Arkansas Rivers. The site where Denver grew up had been a favorite Indian campground, and while miners received threats from the natives to clear off of their land that first winter, a conflict never materialized. The inevitable battle was not averted, but only postponed.

The post–fur-trade pattern of Indian-white relations had been stamped in California where mutual curiosity and friendly exchange was transmuted into unresolvable hostility and mutual genocide. The "mission-civilized" California Indians often went to work in the mines, but their wilder neighbors were pushed aside or run over in the stampede for gold. Both sides believed they had a right to the contested ground, but the whites usually won by sheer force of numbers.

When gold was discovered in Oregon and far northern California in the 1850s, the pattern repeated itself with a new set of tribes, and it became clear that golden ground would be bloody ground. The fact that Colorado remained relatively peaceful for so long was due partly to the concentration of the white population in and near the mountains, where they didn't disturb the plains-moving native Arapaho and Cheyenne. But as farms and road ranches spread east and south from Denver, they offered tempting targets for Indian raiders who still believed in their law of larceny. As poor Robert Stuart learned, what couldn't be held by force was naturally subject to forfeiture.

Outside events leading up to 1864 also contributed to the increase of hostilities. Civil War tensions and persistent rumors of an imminent Confederate invasion heightened the level of belligerence in Colorado. Volunteer regiments were raised to defend the territory, but lacking a Confederate foe, their main actions involved hell-raising in local towns, and stealing livestock to supplement their rations. (Father Joseph Machebeuf's housekeeper, "a strong, well-built Irish woman" named Sara Morahan, caught one party of soldiers raiding her hen house. "She actually seized one of them as he was scaling the fence, and he cried out, 'Oh, let me go, let me go! I haven't got but two!' She let him go, but it was because he tore himself from her grasp.")

At the head of Colorado's illusory defense was John Chivington, a Methodist minister who decided that blood and warfare were more righteous than preaching. The towering, bull-necked parson had distinguished himself in an 1862 defense of New Mexico Territory against an invading Confederate army from Texas. Chivington led a flanking movement that cut off the Confederate supply train, and he burned the loaded wagons and bayoneted five hundred horses and mules to death.

Chivington believed in a policy of extermination where the natives were concerned, an attitude not unpopular with many nervous Coloradoans after the Sioux uprising in Minnesota left more than seven hundred whites dead in 1862. The Minnesota Sioux had been angry because their promised annuity goods did not arrive that summer. They had been mistreated for years and their anger flared into a rampage that led to the killing of every white person they could find. Three hundred Sioux were eventually captured and thirty-nine hanged for the slaughter, but many escaped westward and began inciting their fellow Sioux around Fort Laramie to war and pillage.

When raiding began along the Platte road in the spring of 1864, Chivington ordered his otherwise unoccupied soldiers to pursue any Indians they encountered and to kill them, a policy of shoot-first-and-ask-questions-later that led to a series of bloody skirmishes and reprisals that kept escalating throughout 1864 until the entire mail and supply road along the Platte was paralyzed by Indian attacks.

Food supplies did not reach Denver for more than a month, with shortages becoming serious just at the time local farmers would have been bringing in their crops if they had not fled for their lives. Chivington insisted the Indians were like children, and a little punishment would subdue them. He succeeded only in knocking down the nest and loosing a swarm of hornets.

In a state of panic, Territorial Governor John Evans requested permission to raise a third regiment of volunteers for a hundred-day enlistment, a request granted just when things began settling down in the fall. Evans had sent out a proclamation in June requesting all friendly Indians to congregate at various forts to "protect" them from campaigns against the hostiles. Cheyenne leader Black Kettle, who had demonstrated a long-standing friendship with the whites, led his village of more than a hundred lodges to a campsite near Fort Lyon to await further instructions.

Black Kettle had met with Evans and Chivington in Denver that September, trying to negotiate peace with the whites. Nothing he heard then, or later at Fort Lyon, led him to believe the whites considered him hostile. When Chivington attacked Black Kettle's camp on Sand Creek in the cold dawn of November, 1864, the shock of this duplicity paralyzed the Cheyenne leader, whose only defense was to raise his American flag and a white flag of truce, then run for his life while the soldiers gunned down women and children indiscriminately. Chivington knew this band was peaceful and considered themselves under the protection of the government, but his clear instructions were to take no prisoners.

The battle lasted most of the day, with the soldiers firing into the Indians' defended hiding places, pits dug quickly in the sandy creek bed. The Indian dead were scalped and mutilated for souvenir body parts while survivors fled, dragging their wounded, to a Cheyenne village fifty miles to the east. The Third Regiment of Colorado Volunteers marched proudly back to Denver waving the scalps of their conquest and the citizens cheered this seeming victory. But the reverberations of that day would last more than a decade, like a stone dropped into an unwilling pond, the ripples spreading and consuming the stillness.

An inquiry into Chivington's responsibility for what became known as the "Sand Creek Massacre" led to no more punishment than a tarnished reputation. He maintained to the end that his course had been the correct one, living proof that war, as well as gold, can bring out the worst in a person. The subsequent list of lives lost because of his actions is a long one. They include, on his own side, young Lieutenant Caspar Collins, unlucky Lieutenant Charles B. Stambaugh, arrogant Lt. Colonel William J. Fetterman, and infamous Bvt. General George Armstrong Custer.

"SISTER SYBIL," Hervey Johnson wrote in March of 1865, "as I have nothing to do this morning I thought I would write again. There is mail for us somewhere between here and the Fort [Laramie]. I don't know when it will get here. It is more than a month now since we have had any. Indian troubles are gathering upon us." The fair-faced, twenty-five-year-old Quaker farm boy from Ohio had joined a volunteer cavalry unit to avoid being drafted into the Civil War, and suddenly found himself fifteen hundred miles from home in the middle of an Indian war. It was a war, in 1865, the cavalry was losing. A supply wagon had been attacked by Arapahos the day before, Hervey reported, part way along the lonely, fifty-mile road between Platte Bridge Station and Sweetwater Station on the Oregon Trail telegraph line in Wyoming. Four Indians had ridden up to the wagon on pretense of friendly trade, then shot the guard and barely missed the driver, who whipped the mules around and made a stand behind the wagon until darkness covered his escape. "I was on guard duty last night when he got here; it was after twelve o'clock," Hervey wrote. "He had rode about thirty-five miles without gloves, having lost them in the fight; he was almost froze."

Five soldiers went out the next morning to retrieve the wagon, and found the body of young Private Philip W. Rhoads "stripped of every particle of clothing; he was shot in the left cheek, the ball lodging somewhere in the back of his head, his skull was broken in also by some heavy blow; everything in the wagon was destroyed, except some corn and a box of boots."

Things had been quiet when Hervey first arrived for garrison duty at Sweetwater Station, back in November just before the Sand Creek Massacre. "I scarcely know what to say," he had written then to his sister in Ohio, "but I am so lonesome that I must do something to drive away the 'blues.' I thought when I first came here that I was going to have a nice time, but there was not a stick of wood for winter got up. … For several days past I have had to take the men out to the mountains to get down wood. … One might think it no labor to get wood down from a hillside, but it has to be lifted over rocks and gullies and ravines for hundreds of yards before we can get it where it can be loaded on a wagon."

Hervey declared wood hauling "the hardest work I ever done" until he had to move a stranded supply wagon through snow drifts, cutting blocks of ice like quarried stone from a vertical cornice filling a gulch on the trail. His first view of the valley he would guard for two years had been awe inspiring. "As we passed round the point of a bluff where the road led into the valley of the river, it seemed as if we were going in at one corner of a huge square area that was completely walled in by mountains on the south and west sides. We could see range after range rising above each other, the nearest appearing like huge piles of rock, and as they receded farther back and higher up, everything was lost in snow."

A mile east of Independence Rock, Sweetwater Station was a small log stockade built on a bluff between the emigrant road and the river. It was 150 miles from headquarters at Fort Laramie, and was garrisoned by twenty soldiers whose primary duty was to maintain the telegraph line and keep themselves alive, a difficult task since the army had nearly abandoned the isolated soldiers to the deep, frozen winter of 1865. Rations only occasionally made it to the station, and the soldiers had to hunt wild game or kill a stray ox left behind by emigrants; otherwise it was "bread and coffee and coffee and bread and —— and —— and coffee without sugar and bread without salt soda or grease. I think we lived about two weeks on the above, got nearly down to our fighting weight in the meantime, too," Hervey quipped. "It has not been very cold out here yet," he continued, "the mercury only got as low as sixty-two degrees below zero then bursted the tube." Hervey spent his days chopping through two feet of river ice to get water for the post, and fighting blizzards that swept suddenly across the plains, flinging ice crystals on bitter winds. "There is something sublimely grand in the approach of a mountain snowstorm," he admitted.

The soldiers waited impatiently for the irregular arrival of pay ($18 a month), and mail ("write six or seven times a week," Hervey instructed his family). But the telegraph was their only real lifeline, carrying news from beyond their horizon. Hervey heard about Chivington's attack on the Cheyenne camp a week after it happened. By January he began hearing about the repercussions. Sand Creek had stirred the natives into the action most feared on the western plains, a union of Sioux, Cheyenne and Arapaho forces into a concerted war against the white man. It began at a time of year that usually saw the natives tucked away peacefully in their winter camps, a tornado out of season that struck Colorado first then ripped its way north and west along the emigrant road, leaving death and destruction in its wake.

"The Indians have been committing depredations along the mail road, robbing coaches, and murdering citizens," Hervey informed his sister in January. "Some two or three battles have been fought resulting in the loss of several on both sides. About fifteen hundred Indians were concerned in the affair. They attacked the station of Julesburg, took off the stock, and destroyed everything but the buildings." The Indians returned to Julesburg in February, looting supplies and burning the buildings, then moved northward toward the Black Hills out of reach of the soldiers. Hervey thought it did little good to pursue the Indians. "They watch their opportunity," he accurately summarized, "and when they think there is nobody about or the garrison is weak they light down on the posts, do all the mischief they can, and are gone before men can get ready to follow them."

Colonel Thomas Moonlight, with five hundred troops, pursued phantom Indians from Fort Laramie to the Wind River valley in May without finding so much as a fresh pony track. He came in to Sweetwater Station, Hervey said, and "reported to headquarters that he had driven the Indians all back into the mountains and that it was not probable that they would trouble the road anymore this summer. He had not reached the fort yet on his return till the Indians had attacked several stations, stolen horses and burned one station."

(Moonlight led a later expedition northward into the White River country of the Dakotas, only to have his horses stolen in the night, leaving unhappy soldiers to walk the eighty dry and tortuous miles back to Fort Laramie. Hervey was gleeful at the news. "The old Col. thought himself so brave and smart that it did us good to hear of him having his 'feathers cut.'" Moonlight had severely chastised Hervey's company for losing horses to the Indians, and the turnabout seemed a kind of justice. Moonlight was removed from command after the incident, and mustered out of the army.)

By March of 1865 Hervey was having his own Indian problems. Whirlwinds spun off from the main Indian tornado were rattling stations from Fort Laramie to South Pass, tearing down telegraph line, stealing horse herds, and killing soldiers. When the transcontinental telegraph was built in 1861, it followed the stage and Pony Express route across South Pass on the Oregon Trail. A series of attacks by Shoshones in 1862 impelled the stage line, and much of the emigrant traffic, to move south to the Overland Route, leaving the telegraph isolated and vulnerable. The few forces available, after regular troops were withdrawn to fight in the Civil War, had to be spread thinly between the two roads, easy targets under siege and submerged by the ripple of battle lines that spread ever outward.

Only five soldiers were guarding St. Mary's Station, at the base of Rocky Ridge leading up to South Pass, when Indians attacked there in May. Hervey reported "the telegraph line has been out of repair somewhere for two or three days till last evening, when it was up for about ten minutes." His sergeant tried to send a dispatch to headquarters reporting that half of their twenty mules and horses had just been stolen by Indians, but more important news—that St. Mary's had been attacked and burned—occupied the wire. "A few minutes ago a dead dog came floating by the station. It was recognized by boys here who had been at St. Mary's as a dog that belonged to that station," Hervey shuddered. "He had been killed by the Indians and thrown into the river I suppose, and it is probable that the boys there have met the same if not a worse fate."

Hervey learned later that the boys of St. Mary's were luckier than he supposed. "They had escaped through a rear window when they found the station was on fire and took refuge in an old pit or cellar. The operator attached a wire to his battery and took it in to the cave, with an instrument," telegraphing to Fort Bridger and South Pass Station. "The boys had but one horse at the station and the Indians got him. While they were in their hole one of them saw an old buck riding about on the horse; he shot at him, shooting him through the knee and killing the horse." The incident ended when four hundred cartridges for a Wesson gun began exploding in the burning depot. As Hervey put it, "the Indians thought it wasn't medicine to stay close around and left."

Several Indians appeared at Sweetwater Station not long after the St. Mary's raid. Hervey, on guard duty at the top of the log stockade, sounded the alarm and the horse herd was driven in. The Indians "stood there expecting we would follow them soon," but the soldiers wouldn't budge, wary of an ambush, even when the Indians called them names and began tearing down the telegraph line. "Getting tired they went off, taking with them sixty yards of wire. We watched them as they rode away and it was not more than five minutes after they had started till we could see Indians dropping out of the bluffs ... and in a short time a large party made their appearance and rode off in the same direction."

The soldiers at Sweetwater Station showed good judgment in avoiding the ambush set for them. It was a classic maneuver of Plains warfare used successfully and repeatedly against the cavalry for more than a decade. Caspar Collins wouldn't be so lucky. The frail, twenty-year-old boy slid into his commission as lieutenant on the coat tails of his father, Colonel William O. Collins. Colonel Collins had organized the Ohio Volunteer Cavalry companies that would guard the emigrant and telegraph trails from Fort Laramie to South Pass beginning in 1862. He also established Fort Collins, Colorado, but was discharged from the army just when the Indian war was getting hot in the spring of 1865. Colonel Collins was replaced by Colonel Moonlight, and Moonlight was in turn replaced by a series of hapless commanders whose careers were nearly wrecked by the constant loss of battles in a misunderstood war, of which Second Lieutenant Caspar Collins was only one of the unfortunate casualties.

Young Caspar had lived at Fort Laramie while his father was in command, and was fascinated by the resident Sioux, drawing romantic sketches of their tipis and ponies, and learning sign language so he could talk with them. It must have been trying for him to become an officer and thus an enemy of the people he admired, and frightening to accept responsibility amid such turmoil. "He is a smart scholar," Hervey wrote of young Collins, "but a perfect boy, and entirely incapable of holding the position of Lieutenant."

Lt. Collins was in charge of Hervey's Company G, and made the circuit of his four stations—Sweetwater, Three Crossings, St. Mary's and South Pass—every two weeks. He had ridden close enough to St. Mary's on the night of its attack to see flames arcing into the sky before he turned tail back to safety at Three Crossings. When he rode out from Platte Bridge Station two months later, he wouldn't make it back alive.

The care with which the trap was laid at Platte Bridge is illustrative of how seriously the natives were taking their war. All small raiding parties had been called in, and the tight, silent march of more than a thousand Sioux, Cheyenne and Arapaho warriors to the hills surrounding the station was carried out with the soldiers no wiser. George Bent, the half-Cheyenne son of the old trader William Bent, was along for the show. He had been at Sand Creek when Chivington attacked, barely escaping with a savage bullet wound to the hip. In this mutual war of extermination he was now on the winning side.

Their goal was to decoy part of the post's garrison out into an ambush, dividing the defense so they could storm the station and burn the bridge, a critical link on this white man's road of conquest. In the darkness before dawn on July 26, they filtered down into draws, hid in willows along the river bank, and concealed themselves behind the hills surrounding the station.

Platte Bridge Station had received the temporary reinforcement of various fractional companies traveling back and forth on the road, and 119 soldiers occupied the post on July 26, 1865. Only eighty had rifles, with about twenty rounds of ammunition a piece, another twenty had revolvers, and the rest had no arms at all. The station was expecting a supply train, guarded by twenty soldiers and returning empty from Sweetwater Station that morning.

The soldiers at Platte Bridge had skirmished with more than fifty Indians—scouts from the larger war party—the day before, and it was decided that a party must go to the relief of the supply train in case it was attacked when it neared the station. Lieutenant Caspar Collins was chosen to lead the relief, even though the twenty soldiers detailed to accompany him were not of his regiment. The officers who should have led the party were due to be mustered out soon, and didn't want to risk their scalps.

Caspar might have felt a sense of prescience in this assignment. He had just arrived from Fort Laramie where he had gone to draw more horses for his company. He sent the horses on and lingered at the fort, where he had spent such happy times out riding and hunting and enjoying his boyhood. But General Connor ordered him back to his post, and Collins was afraid he would be killed traveling alone. He arrived at Platte Bridge in the middle of the night, only to hear he must ride out again in the morning. Lieutenant Collins put on his new dress uniform, borrowed a pistol from a fellow officer, selected a headstrong, fine-stepping gray horse, and trotted stiffly away that morning from the station that would become Fort Caspar and eventually the city of Casper, Wyoming. If he couldn't feel a thousand pairs of eyes watching his parade, the soldiers remaining in the station at least were uneasy, sending a foot patrol in his wake to guard the bridge in case of retreat.

Retreat is too orderly a word for the return of Collins's men, when nearly a mile from the post Indians erupted from three sides and mobbed the unwary soldiers. "It appeared as though they sprung up out of the ground," Sergeant Isaac Pennock recounted. "They completely surrounded us. There was no other alternative. … We turned and charged into the thickest of them, drawing our pistols and doing the best we could. It was a terrible ordeal to go through … running the gauntlet for dear life."

Hervey Johnson, writing to his sister, reported the outcome. "Collins was killed while leading a charge. He had charged the Indians once, lost four men killed and seven wounded, and leading the second charge his horse became unmanageable and took him into the midst of the Indians. The last that was seen of him alive, he was riding between two Indians who had hold of him, and both of whom he killed with his revolver. His body was found stripped and so horribly mutilated as to scarcely be recognized."

The Indians had mobbed the soldiers so closely they could not fire without killing each other, and in the wild, hand-to-hand combat, the majority of the soldiers made it back to the bridge where the Sioux, trying to wipe out the foot patrol, had to retreat under heavy fire from their Cheyenne allies who were overshooting at the fleeing relief party.

Barricaded back in the station, the survivors watched as the supply train, unrelieved and unwarned, drove over the hill into a mass of Indians. With barely enough time to corral two of the wagons, the twenty soldiers held out for four hours while the garrison at the station, short of ammunition, horses, and men, helplessly witnessed the slow and total massacre of their comrades.

Unable to do more damage, the large war party drifted back toward Powder River the next day, arguing among themselves over whose fault it was that their plan had not completely succeeded. When General Patrick Connor started three columns (about twenty-five hundred soldiers) to catch and "punish" the hostiles in his midsummer "Powder River Campaign," the cavalry horses died by the hundreds on the seared plains, and his soldiers staggered home again, in rags and starving, harassed by the quick strikes of the natives and defeated by the difficult terrain. The expedition succeeded in killing sixty-three surprised Arapahos in their village, but the main camp of hostiles eluded the army, and General Connor was relieved of command.

The cavalry was losing this war for many reasons, all of which Hervey Johnson was aware. The garrisons were under-manned, poorly supplied and often de-horsed by Indian raids. As Hervey lamented, "I do hope we will be ordered to the Fort or be remounted before long; this thing of belonging to the cavalry and having no horses is getting old with me." Morale was low and the soldiers shaken by their constant defeats, eventually refusing to leave their stockades in groups of less than thirty. The Indian tactic of striking hard and fast where the line was vulnerable, then virtually disappearing when the cavalry pursued, was guerrilla warfare the whites could not contest. As Hervey wrote that July, "the boys out here have all come to the conclusion that fighting Indians is not what it is cracked up to be, especially when it is fighting on the open prairie against five to one. We always have to fight at such a disadvantage, we always have to shoot at them running; they won't stand and let a fellow shoot at them like a white man."

And like another war against natives the U.S. would fight a century later, the goals were uncertain, policies conflicting, and the violence escalated by frustration and misunderstanding. Far from the western frontier, Congress had been preoccupied with the Civil War and divided as to how to solve the western "Indian problem." After each failed military campaign against the natives (and almost all of them did fail) the current commander would be deposed and a peace commission sent to try to buy the Indians' cooperation. Hervey watched the policy change from war to peace to war (and sometimes both at once), and counted the days till he could go home.

Hervey had written with bravado near the beginning of his enlistment that "I have often thought before I became a soldier that I would never try to kill or take the life of anyone, but I have got over that notion now. I could shoot an Indian with as much coolness as I would a dog." It was an attitude born of war, of seeing comrades slain and mutilated, of realizing if you did not kill the enemy, he would kill you. But Hervey finished his enlistment in the summer of 1866 without killing anything (among other things, he was a very bad shot), and he was mustered out, gratefully, just when the next wave of the Indian war was beginning. The army wanted to build a new road through the middle of Sioux hunting grounds to the gold fields of Montana. The Plains Indian war from the beginning had been about gold, from the army's purchase of Fort Laramie in 1849 to protect the California stampede, through the invasion of Cheyenne territory for Colorado gold, then Idaho, and now Montana. And it was a war, in the end, that would be finished by gold.

A YEAR BEFORE THE Indian war had begun in earnest on the plains, prospectors Henry Edgar and Bill Fairweather were guarding their nervous horses in the dark somewhere along the Yellowstone River in what would become Montana. The previous day, May 1, 1863, they had crossed the tracks of a large number of unshod horses, and as dawn lifted that morning, Edgar, Fairweather and their five companions found themselves surrounded by Crow Indians. It was obvious that if they tried to fight they would all be killed, and the seven men watched quietly as the Crows confiscated all their belongings and took them as prisoners back to the Crow village.

They waited tensely the entire day while a Crow council tried to decide their fate. "We talk the matter over," Edgar confided to his diary, "and agree to keep together, and if it has to come to the worst, to fight while life lasts." As night fell Edgar wondered "what will tomorrow bring forth? I write this, [but] will anyone ever see it? Quite dark and such a noise, dogs and drums!" In the morning they were herded into the medicine man's tipi and paraded round and round a bush in the center, then led back outside.

"Bill [Fairweather] says if they take us in again he will pull up that medicine bush and whack the medicine man with it. We tell him not to," Edgar bemoaned, "but he says he will, sure. An order comes again, and we go in and around the bush. At the third time Bill pulls up the bush and Mr. Medicine Man gets it on the head. What a time! Not a word spoken; what deep silence for a few minutes! Out we go and the Indians after us."

Surrounded in the village of 180 lodges, back to back to defend themselves, the white men were saved only by intervention from the chiefs. They spent another uneasy night while their partner, Lou Simmons, who had once lived with the Crows, argued their defense. The Crows were only too aware of the consequences of allowing white men unrestrained access to their territory, so vast half a century earlier and now whittled down to the headwaters of the Yellowstone. At the same time, they didn't want to start a war with the whites that they would inevitably lose.

On the morning of May 4th, the Edgar-Fairweather party was given an ultimatum: continue on down the Yellowstone River and they would be killed; return to where they came from and the Crows would even give them horses and supplies to speed them on their way. "I got a blind-eyed black [pony] and another plug for my [original] three," Edgar complained. The rest of the boys were in the same fix, except for bush-swinging Bill Fairweather who was given his own good horses back, as well as a respectfully wide berth.

Six saddled up and rode away, while the seventh, Lou Simmons, decided it was safer to stay where he was. Not all the Crows thought it was a good idea to turn the white men loose, and Edgar reported being followed as they rode furiously through the nights and hid during daylight until they crossed the Madison fork of the Missouri headwaters and were out of Crow territory. Edgar's group hadn't meant to present such a vulnerable target to the Crows. They'd been hurrying to catch up with a larger party two days ahead, having missed their scheduled rendezvous miles back along the Ruby River in western Montana.

The Yellowstone Expedition of 1863, organized and led by James Stuart, left the new gold-rush hamlet of Bannack on April 9. They waited on the Ruby for a day, then decided Edgar and Fairweather weren't coming. While loitering on the divide between the Ruby and Madison rivers, two members of the expedition found what Stuart called "a splendid prospect on a high bar, but we did not tell the rest of the party for fear of breaking up the expedition." As the goal of the fifteen-man crew was to prospect eastward along the Yellowstone River as far as its junction with the Big Horn, on they went into one of those ironic turns of fate that a man could reflect on all his life.

They crossed the Madison and Gallatin rivers and followed the trail of Lewis and Clark along the Yellowstone. On April 28 they met the same Crows who would capture Edgar and Fairweather four days later. Stuart was invited to sit and smoke with the chiefs. "I complied with their invitation, and our party stood guard over our horses and baggage, while I smoked and exchanged lies with them. It would take me a week to write all that was said, so I forbear. Meanwhile, the other Indians began disputing with each other about who should have our best horses. I requested the chief to make them come out from among the horses and behave themselves, which he did."

Sam Hauser gave a different version of the day's events. "For several hours, in fact during the whole time [Stuart] was smoking and talking with the chiefs, there was a constant struggle and excitement in the camp—the young bucks taking forcible possession of our horses and blankets, and our men by superior strength retaking them, and in many instances handling them without gloves, by throwing them violently to the ground; upon which the Indians would become perfectly frantic with rage, drawing their guns, bows, and knives, pointing to their chief, and making signs that upon a signal from him they intended to take our scalps." A fight was averted only by Stuart maintaining his cool disregard and showing himself the Crows' equal.

When darkness fell the Crows resumed their antics, reaching under the tent sides to grab whatever they could get their hands on. "One thing is certain," Stuart wrote of the Crows, "they can discount all the thieves I ever saw or heard of. . . . They would steal the world-renowned Arabs poor in a single hour." Stuart ordered the horses saddled for an escape at 3:00 A.M., but as Hauser related, "the Indians prevented even bridling our horses." It was clear that a showdown was at hand. "The time had come to die or do," Stuart said. At a prearranged signal the fifteen white men aimed their guns at the Crows, then the Crows dropped their robes and had weapons aimed at the whites. James Stuart leveled his rifle barrel two inches from the chief's heart.

Stuart's "whole features, face, and person had changed," Hauser recalled; "he seemed, and was taller; his usually calm face was all on fire; his quiet, light blue eye was now flashing like an eagle's, and seemingly looking directly through the fierce, and for a time, undaunted savage that stood before him. For several seconds it was doubtful whether the old warrior-chief would cower before his white brother, or meet his fate then and there." The chief backed down and the Yellowstone Expedition rode on to the east.

They passed Pompey's Pillar and found where Captain Clark had carved his name on the rock fifty-seven years earlier. They reached the mouth of the Big Horn on May 5 and found float gold on the sand bars. They had traveled four hundred miles in a month, and intended to prospect up the Big Horn River a ways, then return the way they had come. None of their prospects had been as good as their early discovery in the gulch by Ruby River, but long-standing rumors of gold in the Big Horn Mountains drew them onward. They passed the mouth of the Little Big Horn, caught sight of John Bozeman (who was busy looking for a short-cut trail to the Montana gold fields) but lost him in the brush. (Bozeman thought they were Indians and was running for his life.) And on May 12 they found the tracks of sixteen unshod horses, and realized some of the Crows had been following them.

That night Stuart took the first watch. It was eerily dark and his horses were uneasy. He was lying flat on the ground, trying to see what was out there, when the shooting began, a "terrific volley" of bullets poured into camp at the even darker outline of tents and animals. "I was lying between two of my horses, and both were killed and very nearly fell on me," Stuart related. "Four horses were killed, and five more wounded, while in the tents two men were mortally, two badly, and three more slightly wounded. … I shouted for someone to tear down the tents, to prevent their affording a mark for the murderous Indians a second time. … I then ordered all who were able to take their arms and crawl out from the tents a little way, and lie flat on the ground; and thus we lay until morning, expecting another attack each instant, and determined to sell our lives as dearly as possible."

The spark of their guns would draw return fire, so the determined Crows shot arrows into the camp. "We could hear them whizzing through the air every second, and so near that we often felt the wind," Sam Hauser related. "They were flying in all directions, and it seemed impossible to escape being pierced by them. … so close were the Indians that we could hear the twang of their bow-strings." Hours passed, "hours that seemed like weeks," to Hauser. "Morning came at last, and what a sight it revealed! There was poor [Cyrus] Watkins, shot through the temple and unconscious, but crawling around on his elbows and knees; [Ephraim] Bostwick shot all to pieces, but still alive, and five others wounded." Hauser had been shot in the chest, but a thick memorandum book in his pocket had luckily stopped the bullet.

"On the side of the mountain, in plain sight, were the Indians moving around among the trees and rocks. With the approach of day, the cowardly wretches had quietly retreated up the ravine to the side of the mountain out of danger, yet keeping in sight so as to watch our every movement," Hauser stated. The cool-headed Stuart ordered breakfast as usual. They drank their coffee and pondered their situation. "It then seemed impossible for any of us to escape, but we all had a great desire for some of the party to do so," Hauser said, "and report where, when, and how we had died." In Stuart's reckoning, the best defense was a good offense, and those who were able picked up their guns and went after the Crows. The odds had not improved enough in the Crow's favor, however, and they refused to be drawn into an open battle.

The Yellowstone Expedition packed six days' rations on their remaining horses and set out for the Sweetwater River and Oregon Trail to the south. They left two graves behind. Watkins had died of his head wound, and Bostwick, knowing he couldn't travel and fearing the others would wait with him, put a pistol to his temple and killed himself.

The thirteen remaining men traveled a grueling twenty miles, climbing into the mountains and zigzagging their course to lose the Crows. The next day Henry Geery picked up his gun from the wrong end and accidentally shot himself in the chest. "In spite of our united entreaties, he shortly after blew out his brains so that we could bury him and leave the place before dark," Stuart wrote. "This was the most heartrending scene on the whole trip."

For thirteen days the expedition wandered through the badlands of the Big Horn basin, over the Owl Creek Mountains and into the Wind River valley, dodging the Crows by making a fire at dusk then traveling on into the night. Their six days of rations ran out and they were afraid to fire their guns at game. "It will be a scratch if any of the party are seen any more," Stuart wrote on the fourth day, "but I suppose it is all for the best. Man proposes, and God disposes." On the 22nd of May Stuart thought he recognized the Wind River Mountains. "The appearance of the country has changed for the better. The clay hills are about gone, and we can get feed for our horses. ... Our route, since the massacre, has been through a part of the country too mean for Indians either to live or hunt in, and I came through it to keep out of their way. We are traveling for safety, not comfort."

On the 26th of May the Yellowstone Expedition found wagon tracks, and two days later they hit the Oregon Trail with its telegraph poles, depot stations and emigrant traffic. "Our feelings at seeing the road and telegraph, after running the gauntlet for about four hundred miles through the Crow nation, can be better imagined than described," Stuart sighed. They were back in Bannack by the end of June, cutting north from the Oregon Trail at Fort Hall, having traveled sixteen hundred miles in two-and-a-half months. "Our hair and beards had grown so, and we were so dilapidated generally, that scarcely anyone knew us," Stuart commented. There were few people left in Bannack to see, let alone recognize the adventurers, for one of the ironies of James Stuart's life had played itself out in his absence.

On the same day that Stuart found wagon tracks near the emigrant road, Henry Edgar, Bill Fairweather and their crew had camped on the divide between the Ruby and Madison Rivers because one of their horses was lame. Fairweather (whose name, in another irony, translates to Bonanza in Spanish) called Edgar from his camp chores to help wash out some gravel from behind a piece of rimrock near where Stuart's party had found the "splendid prospect" they would have returned to if the Crows hadn't forced a change of route. Fairweather was hoping to wash out a little tobacco money before they limped on to Bannack, but what he found led Edgar to write in his journal of that night, "a more joyous lot of men never went more contentedly to bed than we." Three pans had yielded two-thirds of an ounce, and the six euphoric prospectors washed eight more ounces the following day, then staked off claims for themselves in what they called Alder Gulch, and lit out for a grubstake at Bannack sixty-five miles away.

Fairweather and company had every intention of keeping their find a secret. But they told a few close friends in "strictest confidence," they spent their grubstake gold a little too freely in Bannack, and when they left town they were followed by a rattling, disorganized, unshakable mob. James Stuart's younger brother Granville, who had stayed in Bannack to keep an eye on the Stuarts' business interests, watched the exodus from the hill above town—about seventy men "strung out for a quarter of a mile; some were on foot carrying a blanket and a few pounds of food on their backs, [and] others were leading packhorses. ... The packs had been hurriedly placed and some had come loose and the frightened animals, running about with blankets flying and pots and pans rattling, had frightened others and the hillside was strewn with camp outfits and grub." Granville helped some of them repack and "soon they were all on their way again hurrying and scurrying lest they get left behind."

By the time James Stuart reached Bannack, the town had been largely deserted, his business made worthless for lack of customers, and the best claims in Alder Gulch were already staked. The Stuart brothers conceded their defeat and moved to Alder Gulch that fall, but garnered little of the $85 million taken out of the gulch and hard rock mines on the hills above. Granville worked as a clerk in the Dance, Stuart and Company Store, while James returned to the wilder country around Deer Lodge.

Never easily stampeded, James and Granville Stuart were the kind of men who gave some stability to the gold rush years—ever watchful for opportunity, never wildly successful, but always getting by. They had been among the first settlers in Montana, and that fact itself resulted from a small turn of fate. Their father, a typical westward-leaning American of Scottish heritage, had moved five times while the boys were growing up, beginning in Virginia and ending in Iowa. In 1849 he joined the stampede to California, returning home in 1851 to take James and Granville back to California. Achieving only moderate success in the far northern mines, the boys, in turn, were headed home in 1857 when they heard that the Mormons (nearly at war with the U.S.) were detaining all Gentiles traveling through Salt Lake. The Stuarts turned north into country that was then part of Washington, later Idaho and finally Montana Territory. They heard rumors of gold in Deer Lodge Valley and there they eventually settled, with more Indian neighbors than white, and just enough gold to keep them interested.

In 1862 a rich placer discovery was made by ex-Colorado miners on Grasshopper Creek, a hundred miles south of the Deer Lodge settlements, and the Stuarts moved down to the new city of Bannack, opening general merchandise, butcher and blacksmith shops. Many who landed in Bannack had been bound for somewhere else. Gold was discovered along the Clearwater River in northern Idaho in 1860, prompting an illegal rush into the Nez Percé reservation where the towns of Pierce, Orofino and Elk City sprang up. The following year saw a rush to the Salmon River, and in 1862 shallow and widespread deposits were found in the Boise Basin. The Idaho rush had come mainly from the west, fed by restless veterans of the California and Oregon mines. Travelers from Colorado and points east who had heard about Idaho gold found the new Montana mines more convenient and often stopped short of their goal.

Bannack City was the major metropolis in Montana (and the territorial capital) until Alder Gulch stole the show. "Varina City" was laid out along Alder Creek in June of 1863, named in honor of the wife of Confederate President Jefferson Davis. Civil War tensions had followed the sixties gold rushes, with hundreds of displaced Southerners crashing like waves onto the generally Unionist western bulkhead. Shortly after the Varina town site was submitted for official approval, as historian Robert Raymer described, "a Republican judge, on being told the name, declared that no such 'rebel' name should besmirch the record of his court and ordered the entry made under the name of Virginia City." Granville Stuart commented that "the name change caused us no little inconvenience. . . . There was already a Virginia City, Nevada, and letters and papers intended for us were sent there and their mail found its way to Virginia City, Montana."

During the first winter in bustling Virginia City, Granville Stuart witnessed the quicksilver nature of gold miners, "a regular stampede craze," he called it. A January freeze had temporarily shut down mining, and "it required no more than one man with an imaginative mind to start half the population off on a wild goose chase." Four hundred dashed away toward the Gallatin River in the middle of January snow, no luck. At the end of January they made for a creek thirty miles from Virginia City. "Men did not stop for horses, blankets, or provisions," Stuart reported; "the sole aim was to get there first and begin to shovel it out at the rate of one hundred to the pan." No *such* luck.

Next the stampeders were off to a branch of the Jefferson. Stuart's partner, Reece Anderson, couldn't resist the urge to join this scramble. He "returned in about two weeks without having found any big thing in the way of gold mines," Stuart chuckled, "but he had accumulated quite a valuable stock of experience and got his nose, ears, and fingers badly frostbitten." The winter weather must have tempered some of the enthusiasm, for the next rush went only as far as downtown Virginia City itself, with lots staked and streets and sidewalks dug up in a dangerous confusion. But the lack of gold became evident before too much damage was done.

Gold hunters like to joke about the time St. Peter wouldn't let a prospector into heaven because too many miners were already in there, tearing up the golden streets.

"Let me in and I'll get rid of them," the prospector said. He turned loose a rumor of a rich strike in Hell and the gates were overrun with stampeding miners. At the tail end was the prospector, who called to a surprised St. Peter on his way out, "You never know, there might be something to it!"

But a certain logic underlay the stampede madness. Late arrivers often found the good claims taken, and their options were reduced to working for day wages or wandering on. Just such a group of late-comers had found little opportunity in Virginia City in the spring of 1864, and pushed on for Canada's Kootenai mines. Returning miners carried discouraging news from that quarter as well, so the four men (two from Georgia, one Californian, and an Englishman) turned back and crossed the Continental Divide into the Prickly Pear Valley fifty miles east of Deer Lodge.

They washed an interesting amount of gold from a little creek at their camp site, but decided to push northward, prospecting up to the Marias River where they found no gold but plenty of grizzly bears. "These they did not desire to annoy," historian Robert Raymer recounted, "for one of their friends the previous winter, as he lay with his cheek bone torn out by the bite of a grizzly, had given them good advice. 'Boys,' he said, 'I've not lost any grizzlies and I ain't huntin' for them anymore.'" The party turned south again and decided to try the little creek they had camped on one more time.

One of the enigmas of gold is that it hides backstage until the final curtain call, on bedrock under ten feet of gravel, or in the dragging tail of a pan full of dirt washed down to a teaspoon. The four wandering prospectors dug holes down to bedrock on either side of what they named Last Chance Creek and saw in the bottom of their pans that they had struck it rich. The town that grew up around Last Chance Gulch would fight with Virginia City for the capital of Montana, and in the end it would win.

Elizabeth Chester Fisk moved to Helena in 1867, seven years before that town wrested the territorial capital away from Virginia City. Twenty-one years old, a newlywed, Lizzie Fisk was from Connecticut. It was a small, hand-written note that ultimately landed her in Helena, pinned in a bundle of blankets she had made for Union soldiers in the Civil War. Robert Fisk was the captain of the New York volunteer regiment that received the blankets. He found the note of encouragement and wrote back to Lizzie in a correspondence that eventually led to their wedding. Lizzie knew that Robert intended to move them to Helena, but she didn't know what Helena would be like.

A three-month steamboat trip up the Missouri brought her to Fort Benton, followed by a thirty-hour stagecoach ride to Helena. Lizzie lost five pounds along the way, "owing chiefly to the want of something to eat. ... Corn and hominy for breakfast, hominy and corn for dinner ... no filter for the water ... only the blackest of brown sugar. ... All this was owing to the stinginess of our [steamboat] captain. I truly believed a meaner man never walked the earth."

Her first view of Helena in July of 1867 was mailed home to her mother. "The streets with few exceptions are narrow and crooked while ditches for mining purposes cross and recross many of them. The buildings are mostly of wood, with occasional stone block relieving the monotony." Building a town on top of mining ground was never a good idea, as the residents of Helena soon realized. Two years before Lizzie arrived, a spate of accidents resulted from the poor planning. A dance hall girl one evening stepped out the back door of the Gayety Saloon for air only to find herself in mid air down a ten-foot mine shaft. Mrs. Combs went into her yard to hang laundry and fell down a fifty-foot hole sunk ten feet from her cabin door. A teamster lost two yoke of oxen in another hole.

In the 1880s a building boom led to the discovery of nuggets in foundation excavations, causing a suspension of construction and brief mining frenzy. Heavy rains still wash gold into the crooked streets of Helena, bringing glitter to the gutters. While the wooden buildings of Lizzie's first view would burn to the ground in one of those devastating fires that swept through almost every mining town in the West, the brick and stone buildings that replaced them would make Helena one of the most picturesque and bizarre illustrations of frontier architecture.

Lizzie quickly concluded that Helena "is not like home. But there is a wide field for usefulness here, and entering upon the work earnestly and prayerfully one need never be lonely or disheartened." Her first work was to make a home for herself and her husband. Robert Fisk was the editor of the local *Herald*, and because the paper was still struggling to pay for itself, Robert and Lizzie had to move in with Robert's brother James, his wife "Goodie," and their fractious little daughter Dell. The four Fisk brothers were widely known as pioneers who led several expeditions from Minnesota to Montana across the northern Mullan Road. While Lizzie liked her extended family, she found her new quarters a little cramped, her sister-in-law a little lazy, and young niece "a contrary little thing." But since she was confiding these observations to her mother, we must allow her to complain.

"For the reason that I am feeling a little blue tonight I ought not to write to you," she told her mother, "but I have a feeling that it will be a relief to talk to someone. ... I have really come to the conclusion that there is never to be, for me and mine anything but this unremitting toil. Robert is working so very hard, he never comes home before one o'clock at night ... [and] when I think how much sewing for myself I have on hand ... there is a silk dress to be made, together with eight pairs of drawers for Dell to be banded, trimmed, &c. ... I want to make some night shirts for Robert, my green dress is untouched, [and] sheets and pillow cases unmade."

Although the cost of living was high in Helena, like most mining towns, Lizzie occasionally gave in to extravagance. "Can you credit my words when I tell you that, in this country where milk well watered is one dollar per gallon, and eggs one dollar and a half per dozen, I made cream pies?" she wrote to her sister. "And today I have been guilty of the further extravagance of cooking for supper a spring chicken for which I paid $1.25 in dust."

By the end of summer, drought had drained all color from the hillsides and hazed the sweeping view across the Prickly Pear Valley. "What would you think of a town with no grass, no trees, no flowers, only dust and stone in the streets and yards?" she wrote home to Connecticut. "Such is our town, and I often close my eyes with a sense of delight as I gaze in imagination on our orchard at home, the dear old grove, the green grass of the yard, with your flowers interspersed among it."

She found the Montana climate trying, scorching hot one day and freezing cold the next. "We have here the most terrible storms of wind and dust which I ever experienced. The dust penetrates every house not plastered, and even then it comes in about the doors and windows." By November it was rain and snow leaking in between the house and the kitchen addition. "The roofs of the two buildings are very imperfectly joined and the rain and snow stream down. The rooms are consequently damp, and, built of rough boards, battened but not lined, are *so* cold."

"Oh, I do so long to get a house of my own," Lizzie pined. But it would be a year before she got her house, moving into one a neighbor was moving out of and leaving Helena for good. In the meantime, Robert departed Helena on a four-month business trip to the East to solicit advertising for his newspaper. Lizzie could not, as a proper lady, leave the house at any length without an escort, and her favorite became a young, handsome doctor from Massachusetts. "I don't know but I shall set the gossips talking," she confessed. "Divorces are fashionable here, and it is a common remark that a man in the mountains cannot keep his wife."

But Robert Fisk kept his wife despite his absences. "Contented, happy wives and mothers are the exception," Lizzie confided. "No one comes here to stay longer than till their fortunes are made, and the idea of building here a comfortable house and making a pleasant home is not for a moment cherished by the majority of our citizens. I tell everyone that I have come expecting to remain here." And remain she did, raising six children in a place where the streets still sparkle after rain, truly a town at the end of the rainbow.

Last Chance Gulch turned out $30 million in half a decade of boom times, and much like Colorado, one discovery led to others, with a total of some five hundred gold-rich streams and gulches washing across western Montana. Most paid wages and little more, but some were true bonanzas. Confederate Gulch, thirty-five miles to the southeast of Helena, produced $10 million. The Butte area, before it became famous for copper, yielded $1.5 million in gold, and the state's total up to 1950 reached nearly 400 million dollars.

James and Granville Stuart, who had been there first and had stayed when others moved on, never caught up with the bonanzas. Both eventually retreated to Deer Lodge, trying to make money at whatever business looked promising in the quiet backwaters. Both served at various times in the territorial legislature. They talked about heading to the gold fields of South America, but in 1872, at the age of forty-two, James Stuart died of liver disease at a lonely, wind-blown agency post on the Fort Peck Indian Reservation.

"We had been together all our lives and passed through many perils unscathed," Granville wrote to his widowed mother back in Iowa. "Our lives were so closely knit together that the separation is dreadful beyond all description to me. I feel like my life was shipwrecked, shattered, and that all our toiling and struggling had been in vain." Granville lived on to see Montana's bonanza shift from gold to cattle, and in 1894 he was appointed four-year envoy and minister to Paraguay, half fulfilling his dream to go to South America. When he died in 1918 at the age of eighty-four he was considered one of Montana's preeminent historians, writing down what he had lived, and he had lived it all.

Of the events the Stuart brothers witnessed during Montana's gold rush years, two, in particular, stand out: the flour riot at Virginia City in the spring of 1865 (which was more a quasi-military appropriation than riot), and the hanging by vigilantes of more than twenty members of an organized band of thieves who were led by the local sheriff. Both events point to the lawlessness endemic to gold rushes and the ultimate human desire to create order amid the chaos.

Montana's remote and difficult terrain had caused supply problems from the beginning. Goods could be shipped by steamboat to Fort Benton, but low water on the upper Missouri often stranded boats short of their goal. A rough wagon road from Salt Lake to Fort Hall and northeast into Montana offered another route for freight, but took months to traverse. Supplies were already short in Virginia City during the winter of 1865 when heavy snows closed the freight roads. A hundred-pound sack of flour that normally sold for $27 rose to $40 by the end of February and continued climbing to $150 for those who could afford it. Most went without, living on a diet of beef (which was plentiful) and grumbling about speculators who had hoarded flour and were now trying to gouge their fellow citizens. Events reached a crisis in April when five hundred armed men marched through Virginia City (their leader on horseback waving an empty flour sack on a pole) and confiscated every pound of hoarded flour in town.

"The search was orderly but very thorough," Granville Stuart observed, "and disclosed sundry lots of flour concealed under coats, in boxes and barrels and under hay stacks," yielding up 125 hundred-pound sacks. It was no act of mob rule, but a disciplined and careful confiscation with receipts duly given for the old price of $27 per hundred weight. The flour was then distributed according to need and all accounts settled, with speculators losing a thousand dollars in the bargain—an imposition of order on what could have been chaos.

Virginia City residents had already demonstrated a willingness to take the law into their own hands. They had been plagued by road agents since the town was founded. No one leaving Bannack or Virginia City was safe in 1863 if he carried gold. Professor Thomas Dimsdale, editor of Virginia City's newspaper, claimed that "one hundred and two people had been certainly killed by those miscreants in various places, and it was believed on the best information, that scores of unfortunates had been murdered and buried, whose remains were never discovered nor their fate definitely ascertained. All that was known was that they started, with greater or less sums of money, for various places and were never heard of again."

The leader of the desperadoes was Henry Plummer, a suave, attractive and reticent man who one contemporary described thus: "With mobile and expressive features he would have been handsome ... a well-cut mouth indicating decision, firmness and intelligence, but not a line of expressive sensuality; a straight nose and well-shaped chin, and cheeks rather narrow and fleshless; still, in their outlines, not unhandsome. But one might as well have looked into the eyes of the dead for some token of a human soul as to have sought it in the light gray orbs of Plummer. ... While other men laughed or pitied or threatened with their eyes, his had the same half-vacant stare, no matter how moving the story or tragic the spectacle."

Plummer had killed two men in California, done time in San Quentin, and left the state a fugitive. Dimsdale claimed that Plummer robbed stages in Idaho before sashaying into Montana to be elected sheriff at Bannack by virtue of his charm. He appointed himself sheriff of Virginia City as well, suggesting politely that the regular sheriff there give up his post, and as Dimsdale reported, "this politic move threw the unfortunate citizens into his hands completely, and by means of his robber deputies—whose legal functions cloaked many a crime—he ruled with a rod of iron." Plummer used his position to collect information on who was leaving town with what, and which stages carried gold worth robbing, sending the word along to his "agents"—a puppeteer pulling everybody's strings.

A group of unsuspecting prominent citizens had Thanksgiving dinner at Plummer's house in November of 1863, the feast including a forty-dollar turkey freighted in from Salt Lake. Plummer was "the soul of hospitality," according to one guest. "His easy flow of conversation, his elegant manners, his gracious attention to his guests made him an ideal host." On an evening in January, without warning or public trial, Plummer was hung by his happy guests and the game was up.

"No law but miner's law" was the rule at the beginning of most gold rushes. An accused criminal was at the mercy of the mercurial mob, the outcome of each hasty trial uncertain. Two of Plummer's dishonest deputies had been set free by a miners' jury after they cold-bloodedly killed an honest deputy who opposed them in June of 1863. But in December the trial of George Ives went the other way. Tall, handsome, blond-headed George Ives was accused of killing a Dutchman named Tiebalt and stealing his gold. Ives, who had survived the grueling Yellowstone Expedition with James Stuart, ended that unhappy year as the first man hung in Montana.

Miner's law eventually gave way to regular law when territories were organized, and officials were appointed or elected. But regular law was not always honest, effective, or just on the frontier, and when crime became intolerable, vigilance committees met secretly to rectify the situation. It had happened in California when San Francisco became unlivable under organized criminal rule, and it happened in Montana on a scale almost unheard of. Including Henry Plummer, vigilantes hanged twenty-one men in a matter of months, and continued their hangings into the next summer. There is always a danger that vigilantes will cross the line between good deed and simple terrorism, from extralegal to unlawful. But it is strange how much more fearsome and remarkable, unsettling and infrequent are acts of vigilance, done for the common good, and how unremarkable and frequent are acts of crime against the common good. Why do we fear one more than the other?

Historians are still arguing about who was guilty and who wasn't in Montana's vigilante episode, but the citizens of Bannack and Virginia City generally agreed that all twenty-one of the unfortunate victims "needed hung." There is doubt, though, about number twenty-two, a man neither murderer nor thief, but a damned public nuisance all the same. J.A. Slade had a reputation for killing any man he didn't like, back when he was a division agent for the Overland Stage line, first at

Julesburg, then Rocky Ridge below South Pass. As Dimsdale wrote of Slade, "he was feared a great deal more, generally, than the Almighty, from Kearny west." Young Sam Clemens, riding the stage to Nevada in 1861, heard all about Slade from the drivers and conductors. Rumor said he'd killed twenty-six desperadoes. He was "a man who awfully avenged all injuries, affronts, insults or slights of whatever kind—on the spot if he could, years afterward if lack of earlier opportunity compelled it," wrote Clemens, who found himself sitting at a stage-stop breakfast table one morning right next to the dangerous man.

"The coffee ran out. At least it was reduced to one tin-cupful, and Slade was about to take it when he saw that my cup was empty. He politely offered to fill it, but although I wanted it, I politely declined. I was afraid he had not killed anybody that morning, and might be needing diversion. But still with firm politeness he insisted on filling my cup, and said I had traveled all night and better deserved it than he—and while he talked he placidly poured the fluid to the last drop. I thanked him and drank it, but it gave me no comfort, for I could not feel sure that he would not be sorry, presently, that he had given it away, and proceed to kill me to distract his thoughts from the loss."

The residents of Virginia City found Slade to be kind and courteous when he was sober, and a "fiend incarnate" when drunk. He would, for example, ride his horse into stores and shoot out the lamps, though he never killed anyone and always offered to pay for damages when he dried out. "It had become quite common when Slade was on a spree," reported Dimsdale, "for the shopkeepers and citizens to close the stores and put out all the lights, being fearful of some outrage at his hands." Dimsdale called Slade's hanging by vigilantes "the protest of society on behalf of social order and the rights of man."

Slade had a devoted wife, who Dimsdale described as "possessed of considerable personal attractions; tall, well-formed, of graceful carriage, pleasing manners, and … an accomplished horsewoman." Warned of her husband's imminent demise, she rode galloping over twelve miles of rough country to try to save him and she arrived too late. It is said that travelers seeking refuge at the ranch where the Slades once lived often awoke in the haunted night to the sound of galloping hoofbeats.

Chapter 9

MAY 13. Snow is falling on the sagebrush buttercups, on the forget-me-nots, on the leafing willows. Two inches last night, another four possible tonight. Hail and thunder this afternoon, drumming on the roof.

Last week I dug prospect holes in Lame Jack Gulch across the ridge. All of the geology looks promising here, with mineral-stained quartz veins cropping out on every ridge, and I chose Lame Jack Gulch for the name alone. I shoveled dirt into gunny sacks from the dry bottom ground (few of the gulches here run water), and hauled the dirt home to wash in my pond. I emptied the sacks across a screen over my homemade sluice box, and dumped buckets of water over the top in a sloshing, muddy, back-straining mess. When I cleaned up the sand from behind the riffles I found nothing. Such work for nothing!

I feel for the tens of thousands who were once similarly employed to no good end. As John Banks commented at the height of the California rush, "the amount of labor thrown away on dams and ditches in California would make fifty or a hundred miles of railroad in Ohio," where he had come from. Only gold could compel such voluntary labor.

MAY 18. It is finally spring, slipping into summer. Buttercups are giving way to yellow violets; the purple-sage colored lupine are just beginning to bud. It gives me great pleasure to notice the variety of species within the monotonical sagebrush. Deceptive it is, such unruly ground cover. One thinks it is the end all.

Spring has eased my sense of isolation. So much seems possible now. There are places to go, and neighbors to gossip about, like the crow that regularly paces my rooftop. It is a most peculiar noise she makes, for she is heavy enough to plant distinct footfalls, a two-legged sound like a hob-nail-booted sprite. She runs lap after lap across the green tar paper, and I imagine it must greatly amuse her.

The Swainson's hawks are back nesting in the aspen grove. Last year they would dive bomb me whenever I came near. This year they think I am a fixture, one more wobbly post along the fence line. Some relationships improve and others deteriorate. I have been in a neighborly argument with the beaver tribe for more than a week. They keep shoring up their dam directly above my bridge. (The original Bureau of Reclamation and Corps of Engineers, their motto is storage, storage, control the flood and flood uncontrollably.)

The bridge was reset on its pilings last June, piece by twisted piece with the help of a come-along tied to my truck bumper, and it is now in danger of washing out again. Every evening I make an assault on the dam, tearing out willow branches packed with mud, trying to lower the water level. But I know the beaver can hear the spilling water as soon as it begins, and they paddle up under cover of darkness to set their world right again. The argument will only be resolved when the spring runoff sinks to a trickle.

MAY 30. Raining. A misty, almost freezing rain, andante on the roof. The sky has lowered to the mountain tops, fingers of mist slipping across and down, reaching out from an east wind. Last night I woke to the sound of snow on the roof, an icy patter that rang an alarm in my mind. I struggled for enough consciousness to care that I would be snowed in here in the morning, but failed. At dawn I looked out the window at the foot-deep drifts, then pulled my head back under the covers. By noon the snow had melted, and this evening it began to rain. The birds, who are always impatient, continue to sing from the drooping willows.

JUNE 13. Sitting on the rooftop, a grandstand seat to watch the evening show. Barn swallows weave in and out below me, their wonderful kite tails and boomerang wings writing Os and Cs and question marks. The nighthawks give their own roaring air show, and gnats whorl around in small, solid tornadoes. All this in still air that smells green like willow leaves and dusky as still water.

Fields of wild iris are beginning to bloom, pale and papery. On the hillsides, bitterroot—all blossom and bud as if the rocks themselves were flowering in unlikely pink and velvety petals. The lupine are now a shocking violet blue, spending their color recklessly. They grow so thickly in places it reminds me of Provence lavender.

This morning I watched blackbirds harassing a crow. They fluttered beneath her large and laboring wings, chattering like squirrels at her raucous screams, another neighborly dispute. Blackbirds are ubiquitous and unassuming creatures, prone to mob rule and plain in appearance unless you look closely, for the male of the common Brewer's blackbird has a blue iridescence on his black feathers, and a piercing eye, dotted black in the center and ringed with gold.

JUNE 20. I have lately returned, tired and footsore, from a prospecting expedition to the eastern hills. I was chasing the rainbow's end to a narrow gulch, indistinguishable from fifteen other gulches, on a drainage of the Wind River accessible only on foot. With three days worth of water and food, camping gear and prospecting kit, my backpack weighed in at fifty-five pounds, an uncomfortable burden for the six-mile hike across hot sagebrush, and nearly my undoing when I came to a tight, five-strand barbwire fence dividing grazing allotments. The pack was too big to slide under or between wires, and too heavy to military press over the top. In the end I put the pack on and climbed the wire at a corner brace, balancing like a tight rope act and imagining my carcass hung upon the barbs, desiccated by the wind like an old cow hide. I cannot deceive you that gold is ever easily won.

I was hiking along a ridgetop, nearly sleepwalking in the high sun and dull sage, when I heard a hissing rattle and saw my dog leap eight feet in the air ahead of me. I never learned to look for snakes, growing up in country too high and cold, and forgot even the possibility, leaving my snake bite first-aid kit at home. I called the dog back with a squeak of adrenaline, and, unharmed, he gave me a look that said "I'm not stupid," and bounced off in search of rabbits and other things that don't bite back.

I inched up on the snake for a closer look—three feet long and two inches thick, "Don't Tread on Me" tattooed on his writhing back. He coiled and raised his head, flicking a forked tongue and shivering his rattle in an unmistakable warning. I backed away and made a wide circle, breathing hard and looking warily under every sagebrush for the last mile into camp.

Despite the dangers and hardships, gold is the best excuse I've come up with to wander around in country other people don't feel a need to see. Gold adds a fourth dimension to landscape, a sense of expectation, a realm of possibility. When I began my prospecting career, I started like a true Forty-niner with only a stamped steel pan and a shovel. From there I graduated to a six-foot sluice box cobbled together with rough lumber and caulk. Next I built a rocker—about the size of an antique baby's cradle—that ingeniously folded up into a backpack. Then I had to have a factory-made aluminum sluice box that I could use in a high-banker (a three-level sorting contraption requiring a power pump), or convert into a rocker when there wasn't much water.

Now I am back to a gold pan, though a slightly improved black plastic model with a fitted sorting screen. I might defend this retreat back to basics by stating that *looking* is infinitely more interesting than *finding*. I am a prospector, after all, not a miner. A prospector travels light, but a miner has to dig in for the long haul. Like most prospectors, I am never satisfied that there is enough gold to make me settle down in any one place and become a miner.

Still, there is a riding hope that I might someday strike it rich, and I was dreaming of gold when I zipped my tent tight that first night in camp and willed it to be snakeproof. There was barely room for the tent on the small, grassy peninsula at the bottom of the gulch, and my dog insisted on coming in where it was safe. If you've ever gone camping with a dog, you probably know that they are terrible tent hogs, and they don't care at all if their feet are muddy. I kicked the dog out at two in the morning to fend for himself, and he was still sulking when I got up. For breakfast I ate a peanut butter and jelly sandwich, which had also been my supper, and studied the prospects for the day.

Gold settles out of moving water in places where the current habitually slows: on inside bends (which is why river bars have historically been so rich), around boulders, across flats and ledges, and in potholes. The dynamics of rushing water can be problematic; currents change with streamflow, riverbeds shift as they snake back and forth across floodplains, uplifts can leave former streambeds high on the hillside, and gold, shaken towards bedrock by the constant turbulence, tends to collect in "pay streaks" accordingly. So it isn't enough just to dig down to bedrock, as many of the early

California fluming companies found out; you have to dig in the right place, a feat that is partly science and largely luck.

I chose a spot where the gulch made a sharp curve around the wall of a bluff, though the fact that my "stream" was less than twelve inches wide made the theory of river dynamics somewhat superfluous. I began sampling from the top down—gold in the grass roots is a sure sign of active deposition—and dumped a shovel full of gravel on the sorter screen. (I learned early on that without a screen you spend hours picking pebbles out of your pan.) Dunking the screen and pan under water, I scrubbed the gravel and shook the pan so anything smaller than the eighth-inch screen holes would fall into the pan. It sometimes pays to look through the leftover gravel for larger nuggets, so I studied the detritus before dumping it on the bank.

Every prospector has his own method for washing a panful of dirt, but the main idea is to shake the pan so lighter material rises while heavier sinks, then to wash out the lighter material with a swish of water. I first dunk the pan completely under water and give it a vigorous side-to-side shaking. Then I start a circle in the water like the beginning of a whirlpool, moving the pan with my arms in an oval motion. I tip the outside edge just enough at the far end of the oval to spin out several tablespoons of material, then tip the inside edge to pick up more water for another stroke. After three to five circles I sink the pan again and shake it, watching the light colored quartz sand rise and keeping a close eye on what is washing out of the pan until only a less mobile wedge of darker sand, a mixture of olivine, garnets, and magnetite, remains.

It takes from five to twenty minutes to wash a full pan, depending on the ratio of light to heavy material. (The more black sand you have, the harder it is to wash down without losing gold.) When I have reduced the pan to that last dark wedge, I empty out all but a tablespoon of water then tip and rock the pan until the black sand walks around the bottom like a show dog and reveals its golden tail. The first pan of the first morning in my little gulch showed only a few golden specks that I washed impatiently away, but in the second pan, glowing like a chip off the sun, was a nugget a third the size of a pencil eraser, and half a dozen smaller colors—the end of the rainbow.

It can be hazardous to succeed first thing in the morning, because you keep after it all day. I was reminded of Henry Bigler, the employee at Sutter's mill who was the first to prospect beyond the tail race discovery site that launched the California rush. He'd gone out hunting one Sunday in February, 1848, but couldn't resist the urge to poke with his knife blade in the crevices along the river. "I sat all the balance of the day in one position, all hunched up, picking [gold] out grain by grain ... and laying it on the top of my cap," Bigler recalled. "The first thing I knew, I could not see. It was dark and being so excited and without thinking ... when I arose to straighten myself, I yelled with pain. A person could have heard me quite a distance. I thought my back was broken. After a few grunts and groans I made my way up the river," and back to the mill, infected with gold fever and stricken with loose lips as well as a sore back.

Like Bigler, I hobbled back to camp with my sample bottle glowing, ate a peanut butter and jelly sandwich (it was all the food I'd packed) and laid my broken back on the hummocky grass. With a high-banker to process more dirt, a person might make a living in that gulch, I thought. On the other hand, looking might be infinitely more interesting (and easier) than finding. I toiled away for another half day just to convince myself that the latter was preferable to the former, then packed up and marched out, nearly stepping on two more rattlesnakes on the way. I've told no one but you about that rainbow gulch, and I'm pretty sure you can keep a secret.

JUNE 24. Once you have seen real gold you can never mistake anything else. I will tell you this cautionary tale so you won't be fooled. My father, as a boy, was out wrangling horses one day with his two older brothers when they spied a mud bank glittering with gold flakes. They waded out in their good boots and dug with their hands, gathering muck in their cowboy hats that sparkled like the grandest treasure. Never mind the horse herd; they could hire someone else to wrangle. What did it matter if their hats were brand new? They could buy more. My father still laughs at himself when he tells the story, but the phrase was already old in Shakespeare's time when he included it in *The Merchant of Venice.* "All that glitters is not gold—Often have you heard that told."

I remember the shock when I washed out my first real gold flake. It was polished and as pure in color as a wedding band. Gold doesn't glitter at all, it glows. And gold doesn't float on the top of mud, or blow away in the wind like sand. In *The Treasure of the Sierra Madre*, where Humphrey Bogart illustrates "what gold does to a man's soul," the ending hinges on the false assumption that gold acts and looks very much like sand. I am telling you now that gold looks nothing like sand. It can run to white if heavily alloyed with silver, or have a rusty tinge from iron, but placer gold always stands out in a dark and weighty crowd of bottom settlers.

The "old Californians" who swarmed across the West after 1850 learned not only the look of gold, but the look of enough gold, a hard concept to grasp for beginners enthralled with any sign of color. James Stuart, arriving gratefully on the Sweetwater after his near-fatal Yellowstone Expedition of 1863, found "plenty of colors" in the gravel bars sixteen miles below Rocky Ridge, but pushed on unsatisfied, searching for a minimum payback of an ounce a day. When they reached Rock Creek, the next stream to my west, Stuart informed his partners that he had found fair prospects there in 1860, "but we are all in too much of a hurry to stop for trifles," he concluded.

That tended to be a frequent conclusion about the gold around here for several decades: there was a sufficient amount to get a prospector interested, but not enough to hold him when other factors interfered—a six-month winter, for example, or unruly Indian neighbors, or bigger strikes in other places. It seemed to be common knowledge among the fur traders at Fort Laramie that there was gold on the Sweetwater. In 1855, before the Colorado and Nevada rushes, the *St. Louis Evening News* reported a party of forty off prospecting the Sweetwater country, but the result of their efforts is unknown.

Disaffected Colorado miners tested the country in 1860 and 1861, only to be driven out by Shoshones. In 1864 Hervey Johnson, guarding the telegraph line at Sweetwater Station, wrote to his sister that "gold and silver have been discovered at South Pass and almost every soldier who was stationed there the past summer has a claim." But escalation of the Indian war, and the constant transfer of company troops precluded any real work on the claims.

Also in 1864, Harriet Loughary, an emigrant to Oregon, encountered a mining camp near Strawberry Creek, downstream from my cabins. "We found here J.D. Jones, once a neighbor in Iowa, who asked us to stop and look over the gold mines. Our obliging Capt. ordered a halt, and as few of us had ever seen gold except in coins, we were glad of the opportunity. Soon all with spades, dishpans, buckets, butcher knives, and wash pans started for the gold, but after some hours of useless toil returned without sight of gold."

In the following three years, various prospecting parties tested the ground on South Pass, and one group finally succeeded in locating a rich lode on Willow Creek in the spring of 1867. Eleven hundred dollars in dust and specimen ore were sent to Salt Lake City for supplies, and as usual, gold in the hand was like fire in the brush. News of a new gold strike along the Sweetwater spread wildly, reaching the Midwest in July of 1867, and topping out across the country. A few set off immediately, and more prepared to join the rush the following spring.

But the newly formed "Shoshone Mining District" suffered an immediate setback when raiding Sioux attacked in July, killing three miners and scattering the rest. Many returned after gathering reinforcements, and several hundred spent the winter in the infant town of South Pass City, battling ten-foot snow drifts and pounding ore in hand mortars. By spring they had crushed $1,600 from the rich "Cariso" discovery lode (the name, through time, metamorphosed into "Carissa," some unconscious bow to lady luck?). They washed another $7,000 from the gulch below, and by early spring of 1868 the stampede had begun.

Gold rush entrepreneurs were now adept at slapping together a city in an instant, arriving just in time to relieve miners of their first flush of wealth in return for any kind of service or merchandise to be desired in the land of Eldorado. They had come a long way from California, had practiced their peddling in Nevada and Colorado, and mastered their art in Montana. Gamblers at heart, they might own part of a lode claim, but they also owned a saloon to hedge their bets, or freighted in an ox train of supplies as soon as snow broke on the trails. Treasure, after all, is where you find it.

And so South Pass City bloomed like wildflowers after a spring rain for sowers who would not stay long enough to reap. Hewn log buildings thrown up in a week were faced with the tall, rectangular false fronts that, freshly painted, were intended to give an air of permanence and respectability. A century later, with paint peeled off and boards warped and weather stained, those same false fronts are painful reminders of the evanescence of human hope. But few were pessimists in the gold rush era—the mother lode was sure to be just over the next ridge, and that ridge dropped into Willow Creek in 1868, if for no other reason than a dearth of discoveries elsewhere.

Merchants were the first businessmen to arrive that spring, erecting large store buildings and warehouses. At C.L. Lightburn's O.K. Store you could buy "groceries, liquors, queensware, dry goods, clothing, hardware, mining tools, stationery, &c., &c. . . . Our motto is Quick Sales and Small Profits," they advertised in the town's new newspaper. Wallpaper could be had at Gaston's, among other essentials, and window curtains at Glenn & Talpey's. Dan Boon (no relation) ran the Auction and Commission Stand, supplying tobacco, cigars, pipes, notions and varieties. "Dan can talk louder, longer and sell more goods at cheaper rates than any other man of his size this side of the Muddy-souri," it was reported.

If you weren't feeling well there were two doctors in town and a dentist. One druggist doubled as a barber, and another sold patent medicines, dye stuffs, paints, oils, glass, and putty; hair, nail, flesh and whitewash brushes; fancy soaps and notions; and ladies and gents braces and trusses. If you were feeling fine you could go eat at the Antelope Restaurant run by a Mrs. Callahan who served up daily "all the comforts and delicacies that our market affords."

At least four hotels were open for business, and also served meals. The hotels were typically two-story buildings, with kitchen, dining room and parlor on the first floor, and two sets of three to four narrow rooms divided by a hallway on the second floor. The rooms were barely furnished with a bed, washstand and water pitcher. The ubiquitous outhouse was downstairs and out the back door. Board at the hotels averaged ten dollars a week in boom times, or more than half an ounce of gold. The City Hotel in South Pass sent a wagon to Utah once a week for fresh fruits, vegetables, butter and eggs, a far cry from the salted mackerel, rusty pork and flour Louise Clappe had to choose from in the California camps.

Two butcher shops supplied the town, selling wild game meat as often as domesticated. You could leave your horse in a number of stables and corrals, or have him shod by one of four blacksmiths. Your mail arrived by pony express—a dollar a letter—and you could ship your gold to Cheyenne with the Nye Forwarding Company, or leave it at the South Pass National Bank which shared a building with the Sweetwater Armory, a supplier of Winchester's new repeating rifles.

But foremost of businesses in South Pass City were the saloons. At least twenty saloons, "snuggeries," and wholesale liquor distributors, as well as several breweries, opened in the town's first three years. Gold rushes had carried with them a cosmopolitan air ever since California's cocktail mix of nationalities, and nowhere was this more evident than in the pretense of saloons. At the Overland Exchange, for example, proprietor Sholes kept "constantly on hand the choicest brands of wines, liquors and cigars that can be found in Wyoming." The opening of the elaborate Magnolia Saloon was a grand affair. "All ye unwashed go to the Magnolia," the town editors admonished, "and take a look at yourselves in that $1,500 mirror behind the long *refreshment* stand, and our word for it, you'll feel a confounded sight better or worse, either one." If music and songs couldn't keep your attention at the Magnolia, their live grizzly bear might.

The Snug Saloon sponsored a boxing match for its customers' amusement. "The performances commenced with a 'setto' between Messrs. Dwyer and Holland, followed by a 'bout' between two gentlemen, novices in the business. Dwyer and his pupil, Mr. Hueston, then exchanged compliments in a style that brought down the house. Mr. Dwyer and Mr. Smith then gave an example of rapid parrying we have seldom seen equaled. We didn't want any in ours," the news editors concluded, "albeit it was friendly." The saloons, as strictly male domains, were often decorated with lively pictures of unclothed ladies. At least one saloon in South Pass kept lively ladies in person.

Despite the encouragement toward recreance and raucousness that so many saloons contributed to, South Pass was a relatively peaceful boom town for a number of reasons. Its citizens immediately sought a regular government, rather than relying on the vagaries of "no law but miners' law." They lobbied the Dakota Territorial Legislature for their own county, complete with commissioners, judges, a sheriff, two deputies and an undersheriff. (By 1869 Wyoming was a territory unto itself, but would struggle for twenty-one years toward statehood.)

Also, the ten thousand hungry gold seekers expected to mob South Pass in 1868 never materialized. Successful gold rushes were founded on rich placers, the kind of gold everyone has access to, washing out of the ground with relative ease and limited capital, a pinch of which would buy a drink at the saloon, and a poke full, a drink for everyone in the house. Most of the gold on South Pass was still stuck in the rock. Fifteen hundred lode mines were staked on South Pass, only a handful of which were developed, but the placer ground proved limited in extent and difficult to work because of water shortages. Counting all comers and goers, the population of South Pass probably never much exceeded several thousand.

The boom might have come larger and longer had there not been better opportunities to the south where the Union Pacific railroad was laying tracks westward at a furious pace, and paying wages to match. Cheyenne, in the southeastern corner of Wyoming, had a population of 10,000 before the rails ever reached it, many of whom, if we can believe the events and eyewitnesses, were not the most solid of citizens. Thus the vast majority of troublemakers and money hunters rushed away from South Pass as fast as they had started towards it.

Among the mad crush at Cheyenne whose original destination had been South Pass was a thirty-year-old Scottish emigrant named James Chisholm. Chisholm had been sent west as a correspondent for the Chicago *Tribune* to cover the much-talked-about Sweetwater Gold Rush, but he found events at Cheyenne more momentous. He had barely stepped off the train in March of 1868 when he witnessed the hanging of two men by vigilantes followed by the murder of another. "The wildest roughs from all parts of the country are congregated here," Chisholm reported, "as one may see by glancing into the numerous dance-houses and gambling hells—men who carry on the trade of robbery openly, and would not scruple to kill a man for ten dollars."

Chisholm was caught for six months in the tumultuous advance of rails across Wyoming, and as a man of refined sensibilities (his past employment had been as drama critic for various Chicago papers), he found the raw violence and corruption in the new boom towns horrific, as the population played leapfrog along the end of the tracks. "The flaring gambling tents—the dance houses—the eternal strumming of old banjoes—the miserable females who have to dance all night till the broad day light, with about as much hilarity as so many prisoners [on a] treadmill—the game of Faro—the game of three card Monte—the game of Roulette, Black and White ... quarrels, cursing, drinking and the flash and bang of pistols," he recorded with awe at the depth of human depravity.

"What will finally become of all that rolling scum which the locomotive seems to blow onward as it presses westward?" he wondered. "Will they get blown clean off the continent at last into the Pacific Ocean?" It was September before Chisholm finally extricated himself from Green River City, the newest end-of-track sensation, to pursue his assignment at South Pass. The reporter caught a ride with a merchant driving a light spring wagon pulled by two horses, a pleasant enough journey until it began to rain and thunder and hail, scaring the horses into a wild runaway in the pitch dark, what the Scotsman called a "mad witch dance on the blasted heath."

Two days in the swaying wagon, and nights wrapped in a soggy buffalo robe on top of squishy mud under the wagon in the pouring rain prompted Chisholm to declare, "such unqualified, unmitigated misery I never endured before." They kept losing the trail and almost plunged over a precipice before finally stumbling into South Pass City where they found their destination practically abandoned, the boom gone bust before it had fairly begun.

"Between South Pass City at three o'clock in the morning and South Pass City at three o'clock in the afternoon, there is but little difference," Chisholm observed. "The actual residents number not over 50 or 60." Most of the local action had temporarily moved east to the Miner's Delight mine and nearby thirty-building settlement of Hamilton City, and to there Chisholm trudged when he had recovered sufficiently from his nights in the rain. He passed through Atlantic City along the banks of Rock Creek where he found "about sixty good log cabins. At first sight one would say, here is a considerable settlement. But when you descend and pass through the silent city, very few of the huts bear any traces of a housewarming."

At Miner's Delight, where specimen ore glowed in the mine shaft, and placer gold spread out like a fan down the timbered slope draining into Beaver Creek, Chisholm found a hard working set of serious men sluicing the better gulches for twenty-five dollars a day. Chisholm reported "no idlers in camp. ... The miners here are a quiet, industrious class of men, mostly old Californians—very intelligent and affording more practical information on mining matters than one can derive from mere book students and theorists." He would later discover that not all of the residents were sober and serious all of the time. "The miners came around in full force this evening [to the store/saloon where Chisholm slept], playing [cards] and drinking to a considerable extent. Perhaps this may afford some explanation of the fact that these men, although earning good wages, are generally moniless ... a vast amount of gold dust is ground in the whisky mill."

As to the feminine residents of the camp, Chisholm stated that "I do not go much into society. Mrs. Gallagher [his host's wife] is my only female acquaintance, and she sometimes pines for home so pinefully that I get quite sympathetic on the subject." The remaining "society" consisted of three other women; the first he described as "a bale of cotton drawn together in the middle, and with a big coal scuttle on the top. She has one white haired little darling and she dotes right onto it. The second is a shadowy secluded kind of a being whose profile I have had a few glimpses of while passing her cabin door. I don't know who or what she dotes onto. The third I will call Dalilah. She is an adventuress. She dotes onto Jack Holbrook. Jack is interested in the Miner's Delight and she is interested in Jack."

At South Pass City, where Chisholm himself went on a spree one Saturday night, he met "a tipsy Bacchanal whose name was Lib—dark eyed—splendid hair but unfortunately drunk. ... She was reciting Shakespeare and passages from Byron and Moore and would have proceeded to tell me the history of her first drunk, but I cut her short. This morning I found Lib sitting at her cabin door playing with a great black mastiff. I sat down upon a rock in front of the hut, when Lib, still half tipsy, set the dog at me for amusement. The brute nearly tore me to pieces."

Chisholm had better luck in the masculine sphere, attaching himself to an experienced miner who was somewhat gruff at first, then warmed to Chisholm's good-natured curiosity. "I tell you sir," Tom Quinn instructed him, "there is but one theory in this world in regard to mines, and it is this—that silver runs in ledges, and gold is where you find it." The fifty-two-year-old Quinn was "a solid looking man of heavy powerful build, slow and deliberate in his movements [with a] meditative aspect and a somewhat stern countenance."

Quinn had been a successful miner in earlier rushes, and had retired to a farm in Missouri to raise a family and quietly pass his remaining years. Yankee soldiers burned him out during the Civil War and Quinn was forced back to the mines to raise another stake. He wanted to see his son educated and his wife taken care of, but he was only making wages, shoveling gravel and swinging a pick for five dollars a day. "He does not play poker nor imbibe whisky," Chisholm reported, "but practices a severe economy," gaining the respect of all. "Which is the bravest," Chisholm pondered, "the man who can lay down his life, or he who can take it up?" We can only conclude this about his success: Quinn was still in Wyoming at the census of 1890, and he died in 1900.

Since the mines and mining towns were lacking in action that fall, Chisholm set off on two excursions into the Wind River valley to the northeast of Miner's Delight, dropping down through the vast canyons and climbing the ridgecaps like a great staircase descending three thousand feet to the basin floor. He marveled at the mountains towering behind him, "silent, untamable, brooding over their own great thought," and poked fun at his own lack of trail skills. "Let an inexperienced man start on a ten day ride without some watchful counselor at his elbow, and ten to one he will arrive at his destination without the coat on his back." Though Mark Twain accidentally set fire to a mountain side while rambling about in Nevada, Chisholm succeeded in setting two huge fires he barely escaped from, as well as losing his pistol, his journal, his horse and his way, a near record in the annals of tenderfootdom.

Settlers had already trickled into the Wind River basin, scratching out small farms and raising stock, unaware that the government had signed the valley away in a treaty with the Shoshone, but very much aware of the danger of marauding Sioux. Chisholm had one brush with Indians on his first excursion, diving into the bushes one way, while the Indians made a dash up a canyon the other way. Chisholm stayed in the brush all day, peeking out through the branches. "I would not wish to insist on it," Chisholm commented, "because one is apt to be mistaken often in things which seem to be palpable facts, but *I think*—I merely put it as a probability—*I think I was a little scared just then.*"

On both of his trips to the valley Chisholm ran into "Mountain Bill" Rhodes, an old hunter with twenty-seven years experience in the mountains who had decided to settle down and farm there, though if too many people followed he figured he'd have to move on. Chisholm was impressed with Mountain Bill. "His life is passed in solitude. He had lived among the Indians, and he has fought them. He knows every range and valley and rock and hollow of all this mountain country as well as 'the oldest citizen' knows his native town." Bill could track and hunt and live off the land, but he decided his life was lacking in one area.

"Bill is not above some of the weaknesses of civilized humanity," Chisholm reported; "he wants a wife." And Bill decided Chisholm was just the person to find one and bring her back; in return Bill would make him a partner in the oil spring he was going to locate. "The future Mrs. Rhodes had got to be a standing joke between us," Chisholm admitted. "I have now got her dimensions, qualities and so forth so accurately noted in my mind that I don't think I can fail to discover her." Chisholm bade adieu to all his friends and left the Sweetwater mines in time to spend Christmas in Chicago with his own sweetheart. Six months later Mountain Bill Rhodes wouldn't need a wife.

"When I traveled this country twenty years ago there was no trouble with the Indians," Rhodes had explained to Chisholm when asked about the recent uprisings. "I have been all through their camps, the Blackfeet, the Siouxs and the Crows, and I turned my horse out and lay down at night with just as little fear as I would in St. Louis or any big city—and a d— sight less. … Now these same Indians, if I got near them, would scalp me. Who has brought about all this change? Why, the government itself."

The government had indeed been busy since Hervey Johnson left the battered army in 1866. The winter of 1865–66 had been brutally cold and long, forcing many of the Sioux to straggle into Fort Laramie seeking provisions and ammunition. The government took advantage of their reduced condition, and sent a peace commission to negotiate for a road through the Powder River basin to Virginia City, a trail pieced together by John Bozeman in the first years of the Montana rush. (Bozeman was a Georgian acquaintance of Green Russell who had arrived in Colorado too late to secure a good claim, and so moved on to Montana. Surviving several close calls trying to open his new road, thirty-one-year-old Bozeman was finally killed by Indians on the Yellowstone River in April of 1867.)

On the heels of the peace commission, who succeeded in gathering the important Sioux leaders at Fort Laramie in the spring of 1866, the army sent a force of seven hundred soldiers to establish forts along the new road, and the Sioux, recognizing the duplicity of this kind of peace by war, vowed to fight to the death for the last of their hunting grounds. Among those who walked out on the treaty makers was a forty-three-year-old, hawk-faced, leading Oglala warrior named Red Cloud.

Red Cloud was born in the White River country of South Dakota when the western fur trade was just beginning. His father died in a drunken brawl when he was a child—the whiskey supplied by Missouri River traders—and Red Cloud received a further education in the white man's wiles when the Oglalas were lured down to the North Platte to trade at a new post on the Laramie Fork in 1835. Red Cloud grew up around Fort Laramie, and when the trading post was sold to the government and fortified in 1849, his troubles began.

First it was the gold rush emigrants crowding through. The Fort Laramie Treaty of 1851 arranged payment to the Sioux for use of the Oregon-California Trail, and in turn the Sioux agreed to leave the emigrants and the neighboring tribes alone. But the agreement was broken on both sides almost before the ink was dry. The Sioux frequently stopped emigrant trains and extorted payment for passage. To the north the Crows, who had been assigned the Powder River country in the treaty, were constantly raided by the Sioux and pushed back toward the Yellowstone. The Shoshone, who traditionally wintered in the Wind River valley, were attacked whenever they crossed South Pass.

On the government's side, annuity payments to the Sioux were frequently siphoned off by dishonest agents, and the garrison soldiers escalated small conflicts by their arrogant attitude. When open war broke out in 1865, Red Cloud joined the hostilities and was vocal in his denunciation of the whites. He had participated in the Battle of Platte Bridge where Caspar Collins was killed, and he would lead the fight against white encroachment into the Powder River and along the Bozeman Trail to the land of gold.

When peace commissioners failed to gain Red Cloud's agreement to a new treaty in 1866, they pretended he wasn't important and declared their mission successful. All would be calm and quiet along the Bozeman Trail they declared, and for some strange reason the military affirmed their declaration. The expectation of peace colored every decision in fortifying the Bozeman Trail, down to the commander in charge, Colonel Henry Carrington, a cautious, detail-oriented man who had shuffled papers during the Civil War. The wives of Carrington and several other officers had been encouraged to go along on this pleasure jaunt, and emigrants to Montana were assured of a safe journey.

In reality, the three new forts along the trail—Fort Reno on the south fork of north-flowing Powder River, Fort Phil Kearny in the middle, 235 miles from Fort Laramie, and C.F. Smith on the Big Horn in Montana—were under siege by July of 1866 while still in the process of construction. Emigrant trains, promised a peaceful passage on that golden road, were surprised by constant and deadly attacks. Colonel Carrington's wife Margaret kept an intelligent and level-headed account of her six months at Fort Phil Kearny.

"July 22d.," she wrote, listing the surrounding events. "At Buffalo springs, on the Dry Fork of Powder River, a citizen train was attacked, having one man killed and another wounded. Indians appeared at Fort Reno, driving off one public mule. Mr. Nye lost four animals near Fort Phil Kearny, and Mr. Axe and Mr. Dixon each had two mules stolen by Indians.

"July 23d. A citizen train was attacked at the Dry Fork of the Cheyenne, and two men were killed. Louis Cheney's train was attacked; one man was killed, and horses, cattle, and private property were sacrificed.

"July 28th. Indians attempted to drive off the public stock at Fort Reno, and failed; but took the cattle of citizen John B. Sloss. Pursuit; recovered them.

"July 29th. A citizen train was attacked at Brown Springs, four miles east of the East Fork of the Cheyenne, and eight men were killed, two were wounded, and one of these died of his wounds. Their grave," she added knowingly, "is still memorial of the confidence with which they left Laramie, assured that all was peace."

Fort Phil Kearny became the favorite target of Red Cloud, isolated in the middle of the long trail, and backed against the foothills of the Big Horn Mountains near present-day Buffalo, Wyoming. It was garrisoned by less than four hundred soldiers who were also carpenters, wood cutters, herders, guards, escorts, and musicians. They were raw recruits, part of the 18th Infantry, untrained to horseback, armed with outdated muzzle-loading Springfields and too short on ammunition for target practice. Margaret Carrington said they did their overly hard work without complaint, but then, she didn't believe in complaining.

The ladies of the fort, who had been encouraged to go along, had their pleasure jaunt cut short. At first they enjoyed horseback rides into the surrounding hills, but then it became too dangerous to leave the prison-like stockade. On September 17, Ridgeway Glover, artist and correspondent for *Frank Leslie's Illustrated Weekly* who was covering events on the Bozeman Trail, was killed a short distance from the fort while out on a Sunday stroll.

"While the ladies could neither ride nor walk beyond the gates, some amusement was attempted between Indian alarms," Margaret explained. "A game of croquet was planned, ... the evening found its recreation in the authors' game, a quiet quadrille, good music, conversation, and other varieties, besides the needle and cookbook." But there were frequent vexations between the few amusements. "When, after a successful trip of six hundred miles, our two cows were driven away one Sunday afternoon by some very mean Indians, there ensued another of those episodes which distract the mind and mar all plans as to butter and cream for cake and coffee. The wolves took our nice turkey hen just as she was ready to give us a brood of little turkeys; while half of our young chickens in that bracing climate gaped themselves to death."

The days were filled with skirmishes and alarms as Red Cloud's warriors attacked wood details, supply trains, and herd guards, but the nights seemed especially ominous. The jumpy soldiers on night guard were prone to shooting at things that weren't Indians. "Sometimes a mule, straying from corral or parting his halter, became the victim of that constant vigilance which was the price of our lives and liberty," Margaret commented; "or sneaking wolves would be mistaken for sneaking Indians, whose habit of borrowing wolf-skins and wolf-cries to deceives us compelled instant attention to whatever had show of life."

Colonel Carrington had taken a defensive position at Phil Kearny, refusing to pursue the marauders, thus drawing the action to within sight of the fort. Margaret had a front row seat, and later apologized for her keen, unwomanly observation and understanding of this war against the natives. "When a woman shares the contingencies of entering a new country with troops, she must learn something besides the lessons of house-wifery, endurance, and patience. When days, weeks, and months pass with constantly recurring opportunities of seeing Indians in small and in large parties dashing at pickets, driving in wood parties, harassing water details, and, with dancing and yelling, challenging the garrison to pursuit, ... when night alarms are common, and three men are shot within thirty yards of the gates, when the stockade becomes a prison-wall, and over its trunks are seen only the signs of precaution or active warfare, ... she acquires *somehow*, whether by instinct or observation, it matters not which, an idea that Indians *will fight*."

Red Cloud would have been proud of her description of how the Sioux did fight. "Dashing directly forward at a run, with the person crouched on the pony's neck, and wheeling only to throw himself out of sight and pass his arrows and bullets under the animal's neck before he returns for a fresh venture; fleeing everywhere, apparently at random, so that his pursuer must take choice of object of quest only to find his hot pursuit fruitless, with gathered numbers in his line of retreat; shooting up and down red buttes, where the horse of the white man breaks down at once; running on foot, with the trotting pony just behind him seeking a rest from the burden of his master; imitating the cry of the wolf and the hoot of the owl when it will hide his night visit, these Indians are everywhere where you suppose they are not; and are certain to be nowhere where you suppose them to be. … In ambush and decoy, *splendid;* in horsemanship, *perfect;* in strategy, *cunning;* in battle, *wary* and careful of life; in victory, jubilant; and in vengeance, fiendish and terrible."

By December it was clear that Red Cloud was serious about his war. During the first six months of occupation along the Bozeman Trail "one hundred and fifty-four citizens and soldiers were killed by the Indians, two score were wounded, and about seven hundred head of horses, cattle and mules were captured; every train going over the road was attacked, or hostile demonstrations were made, with a record of the Indians appearing before Fort Phil Kearny fifty-one times in hostile array," according to historian Grace Hebard. The biggest strike came four days before Christmas.

It was a bright, cold morning, and like many mornings before, the pickets signaled from the hill top that the wood train was under attack yet again. A relief party was immediately organized, and the officer who insisted on leading it would become famous in a way he didn't expect. Brevet Lieutenant Colonel William J. Fetterman had arrived at the fort less than two months before, and was "impatient because the Indians were not summarily punished," for their destructive forays. Fetterman reportedly claimed that with eighty men he could ride through the whole Sioux nation, and that is exactly what he tried to do.

A distinguished veteran of the Civil War's pitched battles and predictable opponents, Fetterman, like many of his post-war colleagues, had little respect for the Indians' fighting ability. He was unaware that the natives, conscious always of their degree of risk, would attack only when they had the advantage, and would fight in a manner that inflicted the most harm with the least damage to their own numbers. As Hervey Johnson had put it, "they won't stand and let a fellow shoot at them like a white man."

Colonel Carrington's last instruction to Fetterman, who badly wanted a decisive battle with the Sioux, was an order *not* to pursue the Indians over distant Lodge Trail Ridge, out of sight and immediate help of the fort. Of course, that's what Fetterman did, decoyed by the retreating Sioux (led by a young warrior named Crazy Horse) over the ridge and into the familiar ambush of fifteen hundred Indians rising up from the ground with bows strung.

Those remaining in the fort listened to the rifle fire. "Every shot could be heard, and there was little doubt that a desperate fight was going on in the valley of Peno Creek beyond the ridge," Margaret Carrington reported. "It seemed long but was hardly twelve minutes before Captain Ten Eyck ... with a relieving party, were moving, on the run, for the scene of action. ... It was just before Captain Ten Eyck's party reached the top of the hill across the Piney, north of the Virginia City road, that all firing ceased." The relief party had arrived too late, and all eighty of Fetterman's command were dead. "The holidays," Margaret wrote of the week following the mass burial, "were as sad as they were cold."

A civilian courier, John "Portugee" Phillips, was sent with dispatches to Fort Laramie begging for reinforcements. "Give me officers and men," Colonel Carrington scribbled. "Only the new Spencer arm should be sent; the Indians are desperate; I spare none and they spare none." Portugee Phillips made the ride in four days of a blinding blizzard, arriving at Fort Laramie near midnight in the middle of the Christmas Ball. The ladies in gowns and officers in dress uniform stood aghast at the man who stumbled in wearing an ice-covered buffalo coat and demanding to see the commanding officer.

When reinforcements finally arrived at Fort Phil Kearny the middle of January, Colonel Henry Carrington was relieved of command. He was ordered to leave immediately for Fort Caspar, and wife Margaret rode away in an ambulance wagon through drifting snow and cold that reached forty degrees below zero. "It was our impulse and duty to go," Margaret wrote, "and we went ... [but] sometimes it seems strange that the trip to Fort Caspar, just then, was such a matter of life or death to the nation, as to make it a question of life or death to us."

By 1867 it was clear, to emigrants at least, that the Bozeman Trail was too dangerous to travel, and the remaining soldiers guarded only themselves. Red Cloud was once again summoned for a peace treaty, but he refused to come in to Fort Laramie until all the soldiers had cleared out of his country. The government decided that the cost of holding the trail had become too high; there were other ways to get to Montana and the gold rush was over there anyway, the Union Pacific Railroad was nearing completion, and nobody cared about the Powder River or the nearby Black Hills. The Bozeman Trail was thus abandoned in the summer of 1868, and Fort Phil Kearny was burned to the ground before the last soldiers were out of sight. Red Cloud, it was conceded, had won his war. The Treaty of 1868 gave the Sioux everything north of the Platte and east of the Big Horns, the telegraph line was moved south to the Overland Trail, the forts from Caspar west across South Pass were abandoned, and the field for the untamed Sioux was left wide open.

Wifeless William Rhodes was one of the first victims of that open field, left stripped and dead in his wagon, his four-horse team cut from their harness in May of 1869. Frank Moorehouse was killed and robbed at his ranch on the Little Popo Agie. On the road to South Pass the Sioux killed "Dutch Henry" Hardyman, and caught Sam Devereux, "a harmless old Frenchman," out planting potatoes. "They surrounded him, beat him down, took the mattock with which he was working, drove it into his stomach, twisting and pulling his intestines out while he apparently was yet alive and begging for mercy," according to South Pass resident H.G. Nickerson.

Shoshone Chief Washakie lost twenty-nine warriors while trying to recover horses the Sioux had stolen. Two more white men were killed in the Wind River valley in August, and by September the Indians were raiding the mines on South Pass, killing John Anderson near Miner's Delight, and a man named Latham on Big Atlantic Gulch.

Seventeen-year-old Frank Irwin set off on a walk from Atlantic City to Miner's Delight the morning of March 31, 1870, only to come upon eight Indians who he thought were friendly until they shot him through the shoulder and knocked him down. "They then advanced to him ... stripped him, kicked and tramped on him, then fired three arrows into his body, two in his chest and one in his left arm, and then started after some other victims, two of whom they met and served in like manner," the *South Pass News* reported. Young Irwin staggered back to Atlantic City, bristling with arrows, and died thirty hours later.

At St. Mary's Station on the old Oregon Trail, Eugene Fosberry, John McGuire and Anson Kellogg were killed. "These men were all terribly mutilated and the place where they were murdered showed evidence of a long and hard struggle," Nickerson described. In June, Oliver Lamoureaux was following the trail of his horses stolen from South Pass when Indians ambushed and killed him.

Dr. R.S. Barr, Harvey Morgan and Jerome Mason were attacked in August near Willow Creek at a place now called Dead Man's Gulch. "These men made a brave stand and hard fight, but were overpowered ... where they could not get to shelter. Morgan was well known to the Indians, having often fed and befriended them, and for his friendship he was mutilated in a horrible manner, the sinews being cut from his back and limbs for bow strings and the queen bolt of his wagon being driven so far into his forehead that it could not be pulled out, but was buried with him as found," Nickerson reported. Morgan's skull was later recovered and displayed at the Lander museum with the bolt still embedded.

The miners on South Pass begged for help from their new territorial governor, and Governor Campbell in turn petitioned the army to come back to the abandoned field. A small force was sent to garrison the new Shoshone reservation (where Washakie refused to come while the Sioux were raiding) in 1869, choosing a site in what is now downtown Lander, Wyoming. In the spring of 1870, another force was detached to South Pass, and their first engagement occurred May 4.

A freight team was under attack near Miner's Delight, and the Indians had run off the cattle when Captain Gordon, Lieutenant Stambaugh and ten men arrived to give pursuit. They lost the trail and were returning when they encountered sixty Indians, and a running fight began. Lieutenant Charles B. Stambaugh rode a faster horse than his companions, and raced right into the Indians' hands. Stambaugh was a veteran of Caspar Collins's and Hervey Johnson's own Company G, Ohio Volunteer Cavalry, and had reenlisted in the regular army when his first term expired. He had come back to the Indian war by choice, twenty-five years old, educated at Notre Dame and raised in a quiet Ohio town where the world was green and well-ordered, not gray and frequently fatal. I don't know what he was thinking.

If he wanted fame and glory, he only achieved it in the most faded fashion. The permanent post established on South Pass between Atlantic City and Miner's Delight was named Camp Stambaugh in his honor. But the post is gone now, only scattered chimney brick and vague depressions mark its location. A creek near where he was killed was also named for him, but it is a small creek on a road seldom traveled. There is gold in the creek bed, washed out of a paleoplacer that spreads across Twin Creek in a boulder-strewn fan, but not nearly enough.

The Indian troubles had begun just when South Pass was experiencing a modest boom in the spring of 1869. Capital was trickling in to develop the lode mines, the transcontinental railroad had finished construction that May, turning loose laborers looking for opportunities, and the population of South Pass swelled to fifteen hundred by summer. Every lot in South Pass City was now occupied by a building or tent, riding, as they thought, on a rising star. But it soon became evident that South Pass was a dangerous place to live, or even go near. Those not scared off by Indians soon became discouraged by dwindling gold returns as the mines deepened and flooded. The garrison at Camp Stambaugh was guarding several hundred instead of several thousand residents by 1872 and the post was abandoned in 1878.

Total gold production from the South Pass mines has been estimated around $2 million, a dim sparkle compared to Alder Gulch's $85 million, or even California's $10 million total in 1849 alone. It makes South Pass hardly worth mentioning in the gold rush chronicles, except for a few outstanding facts. It was Wyoming's largest gold rush, and like rushes elsewhere, it redrew territorial lines, created cities, and brought government to an area that would otherwise have remained unsettled. And as a territory, Wyoming became the first governing body in America to give women the right to vote, fifty-one years before the Nineteenth Amendment made women's suffrage a national law in 1920.

The man who sponsored the suffrage bill in the first territorial legislature was a saloon keeper from South Pass City, encouraged by his neighbor Esther Morris, a six-foot, 180-pound housewife who became the nation's first woman justice of the peace. The press commented on this novelty by describing not what Mrs. Morris did, but how she was dressed. "On the first court day she wore a calico gown, worsted breakfast-shawl, green ribbons in her hair, and green neck-tie," *Frank Leslie's Illustrated Weekly* reported. A bronze bust of Esther Morris still rests in Washington's Capital Rotunda, an obscure reminder of a town that never lived up to its golden promise.

Looking down on South Pass City now, it is hard to believe the town was ever more than the handful of buildings remaining. A photograph taken by William H. Jackson in 1870 shows an extensive grid of streets lined with two-story and false-fronted buildings, and hillsides dotted with sod-roofed miners' shacks, all traces of which have melted back into brush. As a state historic site, what is left of the town is now being carefully restored and exhibited, a sight well worth seeing.

I happened to ride through the town one day on an exploring jaunt. Coming up from the southern canyon on my zebra-legged horse, I must have seemed to appear from nowhere. It was a blustery day and I was wearing a long, black duster and a white hat pulled low in the wind, revolver belted on my hip. The town was deserted except for one family of tourists, and they lined up to stare at me, little children wide-eyed and mother gripping their hands tightly. The father scrambled for his video camera, pointing it uncertainly and stepping backward in retreat. He must have wondered, as I also wondered, if what he saw was real or an apparition, some relic of another time, riding through town on a ghost wind.

Chapter 10

DUST IS BLOWING in my face and stinging my eyes into a Clint Eastwood squint. I pull my hat brim low and duck my head, trying to avoid the worst of the powdery clouds of grit. The wagon ahead of me, creaking and rattling, carries an American flag that snaps smartly in the breeze, and my horse dances and snorts at the whip-like sound. The air smells of dry dust and crushed sage, and the sun is glaring down unfiltered, not a single tree in sight. It's hot, it's July, and I'm playing emigrant on the old, worn ruts of the Oregon Trail.

Our "train" consists of five wagons pulled by mules and draft horses, with six outriders and something like twenty people, most of whom walk because the jolting wagon ride is almost unbearable. It is fifty miles from the eastern flank to the summit of South Pass, and the road here stretches across a barren flat of sand and sagebrush, sand and sagebrush as far as the eye can see. We plod along at two-and-a-half miles an hour, a pace inconceivable in this Ferrari-and-Jetset Epoch. The horizon seems hardly to move and the road passes under us one hoofbeat at a time. In all of this space there is no room for impatience.

The morning wears on slowly, and by mid afternoon we begin to climb into the terraced hills that cradle the Sweetwater River, with green meadows brightening the water's margin. Like a bowl, the hills begin to close in around us, a solid rim ahead except for a crack where the Sweetwater River pours out of its deep canyon to spill across the dry sage plain.

We pass the site of St. Mary's Station, burned by the Sioux in 1865, the small garrison miraculously saved by exploding Wesson cartridges. It was called Rocky Ridge Station in the days of the Pony Express, before the army took it over as a telegraph depot. Noted British travel writer Sir Richard Burton called it Foot of Ridge Station when his stagecoach stopped there in 1860. "A terrible unclean hole," as Burton described it, largely unimpressed with the primitive American West.

The cooking was "atrocious," the floor that was his bed was "knobby," mosquitoes swarmed and wolves "made the night vocal with their choruses." Burton would be happy to know that nothing remains of the station but a marker on a hill above Silver Creek.

Rattling on through the narrowing valley, we find by early evening the triumvirate of emigrant life: good grass, water and fuel in a willow-rimmed meadow at the mouth of Sweetwater Canyon. I unsaddle and picket my horse in the ankle-deep grass while the rest of the camp go about their business, setting up tents and hastily eating their evening meal. Darkness comes early here at the base of the towering ridge, all westward vision eclipsed. No wild wolf chorus vocalizing this night, only quiet stars in an untroubled sky.

In the morning I lead my horse down to the river to drink, then pack my gear and saddle up. One fork of the trail goes into the high-walled canyon, the route John Charles Fremont took in 1842 when he was exploring the Oregon Trail. "We wound, or rather scrambled, our way up the narrow valley for several hours," Fremont wrote. "Wildness and disorder were the character of this scenery. The river had been swollen by the late rains, and came rushing through with an impetuous current, three or four feet deep, and generally twenty yards broad. … On both sides, the granite rocks rose precipitously to the height of … five hundred feet, terminating in jagged and broken pointed peaks; and fragments of fallen rock lay piled up at the foot of the precipices."

John Fremont became the most famous western explorer after Lewis and Clark. The slender, dark-eyed, illegitimate son of a Frenchman had eloped with the daughter of Missouri Senator Thomas Hart Benton in 1841. Jessie Benton was beautiful and seventeen. John Charles was twenty-eight and a dashing lieutenant in the Army Topographical Corps. With Senator Benton's strong backing, Fremont led three official explorations of possible routes to the Pacific, and his well-written reports and careful maps lit like a torch the dimly known lands west of the Missouri.

Fremont climbed to the top of Wyoming's third highest peak, driven by an ambition that occasionally exceeded judgment. The handsome soldier with his beautiful and well-connected wife was chosen to be the first Republican candidate for the presidency, though he lost to

James Buchanan in 1856. His over-zealous involvement in California's Bear Flag Revolt of 1846 had earned him a court-martial, and his privately-sponsored expeditions after he fell out of favor as a government explorer ended in disaster. (A third of his thirty-member party froze to death in the Southern Rockies in 1849.)

Fremont's biographer Ferol Egan commented that "the ultimate curse of being a national hero is that once the fires of acclaim go out, only the ashes of criticism remain." Fremont had climbed to the top, bright, ambitious and daring. He had mapped more of the West than Lewis and Clark. He owned valuable property in California, where he became a senator for one term. He was appointed governor of Arizona Territory, but the citizens there complained so loudly about his frequent absences he had to resign. He lost all his money and property in risky business decisions until his wife Jesse had to support them by writing magazine articles. The harder he tried to be anything but a romantic explorer, the farther he fell from the top of his shining peak. I can only conclude that ambition can take a man a long way down the road, but it can also make him miss the road signs.

The other fork in the trail here begins climbing the massive, stone-encrusted height of Rocky Ridge. "The most crooked, hilly, stony road we have had," wrote an emigrant in 1851. The Oregon Trail to this point has been forward-looking and punctuated by landmarks on the western horizon which become larger and closer until you watch them pass and look forward again to the next monument: Independence Rock to Devil's Gate to Split Rock to the high, snow-skirted Wind River Mountains. But at Rocky Ridge you can no longer see where you're going and you have to climb on faith.

The ridge rises five hundred vertical feet in a mile and a half, and as the hills get steeper, the skinny mule teams begin to stall out, unable to drag their wagons higher. The draft horses are unhitched from their wagons, driven back down the steeps, and rehitched to the mules, towing the whole ensemble upward like a power winch. These were the hills that finally conquered William James in 1856, where John Chislett had to double-team the Mormon handcarts, harnessing extra human power in the drifting snow of October. It is a mountain too hard to climb if you have no more faith.

We finally top out to a view impressive in its breadth, as the Wind River Mountains loom suddenly large. "Nothing that I had before seen of mountain scenery was half so beautiful," Lodisa Frizzell wrote from this point in 1852. "The white snow lying upon the dark ground looked like pictures of silver." South Pass was often a place of superlatives for emigrants of the nineteenth century. It had the steepest hills and the flattest plateaus, some of the best roads and the worst weather, the most spectacular mountain scenery and the least spectacular summit.

One of the true wonders on the old trail was encountering snow on the pass in the middle of summer, a relic of the last receding Little Ice Age. For five hundred years the climate of North America had been in a cooling cycle. Glaciers advanced, precipitation increased, snow lingered and ice stayed frozen in places we no longer find it. At Ice Slough, a noted Oregon Trail landmark between Devil's Gate and Rocky Ridge, emigrants cut ice from beneath the sod where now there is only a melted bog. On South Pass an emigrant of 1851 "found plenty of snow in some ravines by the side of the road. Some of the boys had quite a snow balling." Mark Twain commented here on "the mysterious marvel which all Western untraveled boys have heard of ... but are sure to be astounded at when they see it with their own eyes ... banks of snow in dead summer-time."

From the crest of Rocky Ridge the road drops into an extended meadow with three alkali-rimmed lakes. "Dirty little ponds," Sir Richard Burton called them, "said by those fond of tasting strange things to have somewhat the flavour, as they certainly have the semblance, of soapsuds." The meadow is boggy in places, widely scarred with wheel ruts and littered with leftover props for unsticking stuck wagons. I use the word "meadow" loosely, no luxuriant green swath of beauty, but a rough, hummocky platter of thin grass fighting off the sagebrush.

This is a favorite hangout of the local wild horse herd, a sight unknown to nineteenth century travelers when horses were too valuable to allow free range. But the emigrants often noted the animals they did see on South Pass, a list that included prairie dogs, ravens, grouse, elk, antelope, a few buffalo, deer, wolves, black bears, and clouds of mosquitoes. Prairie dogs are rare here now, but I have seen deer, elk, antelope and even buffalo on the pass. Absent from last century's list were moose (which have gradually worked their way southward from Canada), a startling, incongruous shape lumbering across the sage.

At the far end of the meadow we cross another ridge and drop down to Radium Springs, with two buildings still standing amid the debris of others, the wreckage of Old Lady Gillespie's dream of empire. I stop for a long drink of cold water, then ride on, plodding hoofbeats laying tracks on the sunken road. Over the next ridge is Strawberry Creek and the Lewiston Mining District, a near war zone of prospect holes and mine shafts, heaped tailings piles and leaning headframes. The names of the abandoned mines speak volumes: the Hidden Hand, Bullion, Good Hope, Mint, Gold Leaf, Big Nugget.

Strawberry Creek was a favorite campground for the emigrants after the long pull up Rocky Ridge. "It supplies plenty of the fragrant wild fruit," Sir Richard Burton wrote in 1860, "and white and red willows fringe the bed as long as it retains its individuality." When Lodisa Frizzell reached Strawberry Creek in 1852, she realized she was not even half way on her journey. "I would have given all my interest in California," she wrote, "to have been seated around my own fireside, surrounded by friend and relation. That this journey is tiresome, no one will doubt; that it is perilous, the deaths of many testify, and the heart has a thousand misgivings."

We press on into the afternoon, passing my cabins tucked into the base of a tall ridge. The mules' ears droop and the harness chains clink with every step. The "wind blows hard & clouds of dust nearly blind one," Peter Decker wrote in 1849. At last we reach the murmuring water of Rock Creek, our campsite for the night. My unsaddled horse rolls in the short grass, scrubbing off the dust and sweat. By nightfall a fiddle and guitar appear amid the group, their sawing music urging on a square dance round the blazing campfire. I seem to be the only one aware that we are dancing on the grave of William James. No act of disrespect, it is merely a forgetfulness, born with every generation, a *tabula rasa* which we must fill as we see fit.

Morning is preceded by a rainstorm, cold and blustery. It requires fortitude to crawl out of my snug mountain tent into the damp chill. I am met with dour faces; the men scratch at several days' growth of beard and the women have lost the hairstyle they began with. As a game, this is losing its amusement. We straggle out onto the trail by late morning, across flat, sage-covered terrain, three miles to Willow Creek. The crossing at Willow Creek is boggy, and the wagon drivers debate for half an hour whether the mules will stick fast in the mud. All hands to the wagons, they finally decide, and the brave new emigrants wet their feet in the stream as they push the wagons across while the mules lunge and the drivers curse and slap the reins.

We climb again to a high, flat stretch, the wind gusting cold, and heavy clouds scuttling eastward. The weather was often unseasonable for travelers leading the year's migration, as William Johnston commented in early June of 1849. "For inclemency this day exceeded any since our journey began. From early dawn the sky had been frowning with black clouds, and a thick mist pervaded the atmosphere. Chilling showers of rain and sleet followed, driven furiously by piercing winds, so that in spite of a superabundance of heavy clothing we were nearly frozen." I pull my own oilskin duster tight as the trail drops down to the last crossing of the Sweetwater.

The mountain men traders accrued good profits here from the California Argonauts, selling "fresh" oxen they had taken in trade a week or two before, shoeing mules and proffering that most western of accessories, the buffalo robe. The Mormons found the last crossing a handy site for a mail station in 1857. A man named Gilbert kept his own station opposite the Mormons, and here Colonel Frederick Lander, building a shortcut to the north of the Oregon Trail, left Charles Miller to record the winter weather on South Pass. It snowed November 2, and the wind blew hard and often. In December, mail arrived from both directions, the carriers' "feet, hands and faces frozen badly." January was blessed with a few pleasant days and on February 4, 1859, this notation: "Observer killed and buried." The hangers-on at Gilbert's Station were Southerners, and in a war before the war, it seems, Charles Miller was a Yankee.

Mark Twain rolled in to the stage station here and met the Allen's revolver of dignitaries. Sir Richard Burton found it necessary to "liquor up" at South Pass Station "with a whiskey which did not poison us, and that is about all that I can say for it." Soldiers from this station rode to the rescue of the St. Mary's survivors, and like St. Mary's, South Pass Station was set afire by Indians, and has since been known as Burnt Ranch.

Wagons forded the Sweetwater River here for the ninth and last time, climbing a steep bluff to the final, nearly level pull across the dividing line between the Atlantic and the Pacific at 7,550 feet. "The road is as broad & fine as any turnpike," wrote an emigrant of 1850. "The ascent was so smooth and gentle, and the level ground at the summit so much like a prairie region, that it was not easy to tell when we had reached the exact line of the divide," another commented. A stone monument marks the spot for today's travelers, not quite half way on the long road to Oregon.

It feels unpatriotic to turn around here, to head east against the western swell of manifest destiny, but so I must, joining the ranks of the disaffected, the unfortunate, the washouts. In my defense I can name some notable circle makers: Lewis and Clark, Pike and Long, Powell, Fremont, Hayden. Government-sponsored explorations were generally famous for their elliptical swaths of information gathering, but one, at least, was infamous. In a circle that hardly warrants a footnote in the annals of great explorations—no grandly dangerous continental crossing, no one-armed ride down a wild river—it was a bureaucratic chess move on the Dakota prairie to document a small corner of land that had been too troublesome to bother with in the past.

The upper Missouri River had become settled in the early 1870s by Americans restless for more ground to overrun. To their west lay an isolated dome of mountains rising up from the plains in a promise, skirted by rumor and innuendo, that the deep and tangled canyons there were paved in gold. They'd go see for themselves, but prospectors who made the mythic journey to the Black Hills tended not to come back. It was part of the territory assigned to the Sioux in the Fort Laramie Treaty of 1868, and the Sioux took exception to trespassers.

From the beginning, the army's decision to explore the Black Hills in 1874 was marked by contradiction if not outright duplicity. The official order authorizing the expedition blandly stated that its purpose was to "examine the country" from the Belle Fourche south to the Black Hills. Frontier communities interpreted this blank check as the first step in opening the land to settlement and exploitation. Indian advocates called it "high-handed outrage," and predicted Sioux retaliation. Certain members of the military hoped it would lead to a fort in the middle of Sioux territory to better control the wild natives.

To further muddy the waters, the original order authorized only an engineer to map the area, but the commanding officer of the expedition surreptitiously added a geologist, paleontologist, photographer, four newspaper correspondents, and two civilian prospectors, not to mention then-President Grant's drunken son. The commander himself was a body of contradictions. Thirty-four years old, slender and fine-featured, he had graduated last in his class from West Point, and was court-marshaled twice in his career. He was married to the beautiful daughter of a judge, who encouraged his temperance and refinement. He had distinguished himself during the Civil War with brilliant leadership in bold and desperate battles against overwhelming odds, achieving a brevet general's rank. Though he was only a lieutenant colonel in the 7th Cavalry in 1874, everyone but his superiors called him "the General."

"I came here expecting to find a big-whiskered, swearing, ranting, drinking trooper," wrote the correspondent for the *Chicago Inter-Ocean*. "I found instead a slender, quiet gentleman, with a face as fair as a girl's and manners as gentle and courtly as the traditional prince." Others were not so generous in their assessment of the General. Private Theodore Ewert, a trumpeter for the 7th Cavalry, suspected his commander was seeking self-glorification above all else. "The hardships and danger to his men, as well as the probable loss of life were worthy of but little consideration." Indeed, Ewert's suspicions would prove prescient. Brevet General George Armstrong Custer was a man of grand ambition where glory was concerned.

Custer's final act of ambiguity as leader of the innocuous 1874 expedition to the Black Hills was to send messages of peace to all of the Sioux bands, then march out of Fort Abraham Lincoln (across the Missouri from the new town of Bismarck, North Dakota) on July 2nd, with over a thousand men armed to the teeth and ready for war. We could interpret this act as correspondent A.B. Donaldson did: "The very strength and formidableness of preparation are to enforce a peace by rendering opposition futile and disastrous." But never before had a peaceful mission of exploration been so war-like.

Sixty Indian scouts formed the vanguard of the expedition, followed by the General and his staff. Eight companies of cavalry flanked the column of 115 supply wagons, ambulances, three Gatling guns capable of firing 350 shots per minute, and a Rodman artillery piece guarded by two companies of infantry. Well over a thousand horses and mules pounded a trail across the prairie visible for decades by the sunflowers growing on the disturbed sod.

Five of the infantry collapsed the first day out on the blistering, grasshopper-bitten, rolling swells of prairie. The horses' hocks, pricked by cactus spines, left a bloody trail in the sparse grass. Wood was scarce and water alkaline, but the General was having a grand time out hunting on his fine horse, like European nobility of decades past. "Every day the General may be seen in buckskin and broad-brimmed hat, accompanied by his faithful hounds," noted the *Bismarck Tribune* correspondent. No one but Custer was supposed to enjoy the sport of hunting; he angrily tried to crease the heads of two scouts who downed an antelope within his sights.

But antelope presented too much of a temptation even for the enlisted men. Correspondent Donaldson described how a herd "ran right across the advancing columns within ten or fifteen feet of the battery of artillery ... and whole volleys were discharged contrary to orders." All but one antelope miraculously escaped this onslaught, but the show wasn't over. "The driver of one of the four-horse teams drawing a Gatling gun left his post for a moment to salute the herd with his six-shooter. His team became frightened and ran away, and for a few minutes a Gatling gun made a series of rapid evolutions over the plains of Dakota, in a manner wholly unrecognized in military practice. Before any damage was done, the war steeds ingloriously terminated their brilliant maneuver by miring in a slough. The driver escaped with no other punishment than that of being dismounted for the remainder of the day."

Twenty days and 350 miles out, Custer's expedition finally reached the Black Hills, an oasis of cool timber and beautiful park lands. Custer entertained himself by climbing the highest of the nearby peaks and engraving his name at the summits, while his civilian prospectors rooted around in the streambeds for gold. On French Creek, near present-day Custer, South Dakota, they finally found pay dirt "from the grass roots down." In truth, the diggings were marginal, an optimistic ten dollars a day per man which sounded good back in the States if one neglected to factor in the cost of transportation, supplies, and a probable Sioux hair cut.

It was gold enough, though, to wreck the soldiers' discipline while Custer was away on a exploring jaunt. A day after the civilian prospectors brought back a pile of golden grains wrapped carefully in a sheet of account book paper, the *Chicago Inter Ocean* correspondent witnessed a sudden crowd around the new diggings "with every conceivable accoutrement. Shovels and spades, picks, axes, tent-pins, pot hooks, bowie knives, mess pans, bottles, plates, platters, tin cups, and everything within reach that could either lift dirt or hold it was put into service by the worshipers of that god, gold. And those were few who didn't get a 'showing'—a few yellow particles clinging to a globule of mercury that rolled indifferently in and out of the sand."

Custer, himself, was above such antics, though he recognized the value of the discovery in furthering his own ends. While his official report would appear cautionary and disinterested, he sent a galloping courier off to the telegraph station at Fort Laramie with news dispatches of a new gold strike. "Until further examination is made regarding the richness of gold, no opinion should be formed," Custer officially warned. "GOLD AND SILVER IN IMMENSE QUANTITIES," screamed the headline of the *Bismarck Tribune*. Back in June that same newspaper had speculated that "Custer's expedition may be the pebble which, dropped in at an opportune moment, will set the mighty sea of American thought in motion." More than ten years had passed since a really rich gold strike had drawn crowds westward. A farm and financial depression in 1873 was pushing hard at people who wanted only the small pull of a new Eldorado.

So Custer marched back to Fort Abraham Lincoln wearing a golden halo of glory as the man who had transmuted rumor into fact. He had climbed his mountains, killed his first grizzly ("the hunter's highest round of fame"), and lost four men, three to dysentery and neglect, and one to unprosecuted manslaughter. As for Indians, the showing was poor. Custer surprised a small mountain tribe in the Hills, whose old chief was held captive as a guide while the rest fled for their lives. "The dusky natives have been seen in small bands, but they have not interfered with us," noted correspondent Knappen. "We have not even had a good scare," he complained. Still, Custer had indeed succeeded in dropping the pebble.

Back in the States, Custer's superiors scrambled to dampen the waves Custer had raised. Prospecting companies were already organizing for a fall invasion, and while everyone recognized that the Black Hills were part of the Sioux reservation, they did not at all conclude that it was Sioux gold. General Sheridan, commander of the Division of the Missouri, ordered out patrols to watch for trespassers, and gave permission to burn wagons and arrest the leaders of any expedition en route to the Hills. The frontier fire was thus somewhat cooled, but one party managed to slip through the army's net, leaving Sioux City in October with six wagons, and claiming the O'Neill Colony in Nebraska as their false destination.

Twenty-six men, one woman and a boy, "the first expedition to the Black Hills, cut loose its prairie craft from its moorings on the banks of the 'Big Muddy,' and followed the 'Star of Empire' westward, right through the heart of the Sioux reserve," Annie Tallent recalled. A self-possessed, whip-like woman, Tallent accompanied her husband and nine-year-old son on the expedition, walking almost every mile of the way.

They were all light-hearted and adventurous at first, crossing the Nebraska prairie from settlement to scattered settlement, camping out "under heaven's dark canopy, with its myriads of bright stars twinkling lovingly down upon us like a very benediction." They told stories and sang songs in the evening, a comradeship among virtual strangers for they were gathered by purpose, not familiarity. "I must confess," Annie Tallent wrote, "that I really enjoyed those social hours spent around the smoldering camp fire after our days' journeys were ended."

Those first weeks of easy travel must have glown even brighter amid the darkness to come. Half way to O'Neill one member turned back, claiming sudden and severe illness. "The poor fellow just became 'awfully' homesick," in Tallent's opinion. Twenty-five men, one woman and a boy pushed on to O'Neill where the members of that remote colony tried to dissuade Tallent from going on. "All their well-meant advice went for naught. … After a day spent in the O'Neill settlement for rest, our journey westward was resumed … how utterly horrified those kind people looked as our train pulled out of camp."

They were now, as Tallent put it, "treading on forbidden ground. … At any time we were liable to be met or overtaken by roving bands of Indians, who we felt sure would look with no favor upon our aggressive movements. On the other hand, we were still more afraid of the authorities we had secretly defied. We were in constant expectation of seeing a troop of cavalry come upon us from the rear, seize our train, burn our wagons and supplies, march us back in disgrace, and possibly place us in durance vile."

They made wide, gyrating detours to confuse trackers as they crossed the plains northwestward to the Niobrara, Keya Paha and White rivers. Water became scarce, and frozen blocks cut from White River were carried in a wagon and melted for drinking. "The water thus secured was in a high degree offensive and nauseating, wholly unfit for man or beast, and not until nearly famished with thirst could I be tempted to drink a drop of the vile compound … [an] unpalatable conglomeration of chalk and congealed water," Tallent declared.

One of the party died of dysentery from drinking the bad water, his groaning cries from a wagon bed finally silenced. They buried him in a hewn log coffin to keep out the wolves, and placed at his head a hewn log cross to keep out the Indians. "Gloom, like a dark pall, hung over our little camp on the dreary, lonely prairie that night. Death was in our midst and every gust of wind that blew down the valley seemed laden with the wails and groans of our departed companion."

Their oxen were growing weak as they reached the South Dakota Badlands with its weirdly crenelated formations that resembled, Tallent thought, "fortresses, castles, and even small villages." Leather boots froze solid in the November cold, and they had to wrap gunny sacks around their feet and legs. Camp fires were put out before dusk, so neither Indian nor soldiers would see the points of light in the darkness. Tempers grew short and guns were drawn but not quite fired. "A trip over the plains with all its trying discomforts brings to the surface the most unlovely elements of a man's character, or a woman's either for that matter," Tallent concluded.

Across the Badlands to the Cheyenne River, the group plodded on toward the Black Hills. "Let none of my readers be deluded into the belief that there was anything, either very romantic or pleasant connected with this part of our journey," Mrs. Tallent asserted, "unless shivering over the dying embers of a campfire, silently watching the daylight gradually fade into darkness, until all the surrounding desolation was overspread with sable wings of night, and then creeping, benumbed with cold, into bed, be romantic, or unless getting up at early dawn, partaking of a hastily prepared breakfast, not too tempting to the appetite, and trudging off through the snow, day after day, be considered a pleasure."

It was December before they caught sight of their goal, "bold, rugged, abrupt mountains." They found Custers's trail and followed it toward French Creek. "All along the route could be seen in places, on one hand, huge rocks piled high one upon the other, with almost mechanical regularity and precision, as if placed there by the hand of a master workman—a great wall of natural masonry; on the other the everlasting hills, covered with majestic pines that looked like stately sentinels guarding the valleys below, towering far, far up above our heads; then anon low lying ranges of hills, clothed with dense forests of pine, and away in the hazy distance, other ranges rising up like great banks of clouds against the horizon."

The group arrived at French Creek, two miles below the present site of Custer, on December 23, 1874, seventy-eight days after they left Sioux City. On Christmas Day there were no presents, even for the boy, no "roast turkey with cranberry sauce, plum puddings, and mince pies," and the day after Christmas it began to snow, "coming down in great feathery flakes until the whole landscape was covered to a depth of two or more feet, on a dead level, and our tents were almost literally snowed under."

In January the men constructed an eighty-square-foot stockade of upright logs set three feet into the ground to guard against an Indian attack that never materialized. Inside the stockade seven cabins were built along the walls like a small city of strangers. Each separate group built according to their abilities and did not help the rest. Thus one cabin was made by "a half-dozen fine muscular fellows from the pineries of Wisconsin who were not afraid of work." Their cabin "was conspicuous because of the peculiar construction of the roof, which consisted of small hewn timbers with a groove chiseled out in the center of each to carry off the water. As a substitute for shingles it was an ingenious contrivance. This same cabin had a floor of hewn logs, a door of hand-sawn boards, a chimney, a fire place, and an opening for a window, but no sash."

Annie Tallent's cabin, on the other hand, was the "most unpretentious of the seven," a one-room shed of unhewn logs with a dirt floor, pole and branch roof, gunny sack door, cockeyed fireplace, and an adjoined lean-to occupied by a hot-tempered Moor. A small window connected the cabin and lean-to, through which would pass daily courtesies and borrowed implements. Mrs. Tallent had a small iron kettle which she frequently loaned to Mr. Cordeiro, and he owned a razor-sharp double-bit axe. "As these implements were being passed back and forth through this convenient aperture, our neighbor, when looking through from his little dingy room with his supernaturally intense black eyes, made a very suggestive picture."

The miners managed to dig only forty dollars worth of gold from the frozen ground, but it was enough, they thought, to start a spring rush in Sioux City. Two men left the sixth of February, taking the gold and three horses away towards civilization. Two more left the middle of February with an ox and a sleigh, escaping to Fort Laramie. In March four more defected. "Two of the deserters, having saddle horses, rode away with blankets strapped onto their saddles behind, and guns across the pommels in front." Another led his possessions away on a donkey, and the fourth walked away with a gun across his shoulder and a pack on his back.

For Annie Tallent it was a long, unhappy winter. "Imagine yourself imprisoned within the gloomy walls of [a stockade] enclosure, and more closely confined within the still gloomier walls of a cell-like cabin, with no work for mind or hand to do, and with an uncertain fate hanging over your head, and you may be able to form a faint conception of the misery of life in the old stockade during the memorable winter of 1874–5."

By April, the army had located the expedition's stockade, and gave the remaining inmates twenty-four hours to pack their bags. Mrs. Tallent rode away from French Creek on the back of a government mule, reaching Fort Laramie in ten days where the army set them loose. "We were back again within the pale of civilization and the law, after an absence of nearly seven months. Thus ended the memorable journey in and out of the Black Hills, with its dangers and hardships ... the members of which gained nothing save a dearly-bought experience."

Other miners were slipping into the Hills even while Tallent's party was escorted out. It was a game of cat and mouse, and the mice were winning. By August, 1875, some six hundred men had made their way into the Hills, and General George Crook was sent in to remove them, an initiative to mollify the Sioux for upcoming negotiations. The government was willing to buy the Black Hills, but the Sioux, it turns out, weren't willing to sell.

When nine "peace commissioners" met with some twenty thousand Sioux in September of 1875 at Red Cloud Agency, they found the mood tense and volatile. The Sioux could feel the noose tightening—their hunting grounds eroded, buffalo disappearing, treaty provisions ignored and white men gaining ground on every side. They would not give in again easily. Surrounded by unhappy Indians, the commissioners tried to bargain while warriors thundered down on them firing into the air and threatening to kill any chief who would sell the Hills. Red Cloud, sullen and demanding, wanted his people supplied with all the white man's goods (down to four-poster beds in regular houses) for seven generations. The other chiefs arrived at a price of $70 million, $64 million more than the government was willing to pay. Fearing for their lives, the commissioners declared the negotiations a failure and fled back to Fort Laramie.

The breakdown of negotiations convinced President Grant, in consultation with his secretaries and generals, to rethink his position on the Black Hills. While the area would still legally belong to the Sioux, the army would no longer interfere with trespassers. Thus the gates were effectively thrown open to any who wanted to assume the risks. Miners poured back into the Hills, with French Creek and the new town of Custer as their destination.

Had French Creek been the sole source of gold in the Black Hills, the rush would have been brief and excoriated. Much like the Pike's Peak Rush, where Denver was optimistically founded on the thin gold of Cherry Creek, it would require more and bigger discoveries to carry the rush beyond a ruse. Prospectors spread north from French Creek in the fall and winter of 1875, making small strikes in dozens of gulches and stream beds. They would find evidence of earlier gold hunters—old cabins in overgrown forest, mining tools at the bottom of slumped prospect holes—who hadn't lived to tell their treasure tales.

The most enigmatic piece of evidence was a slab of sandstone found by Louis Thoen in 1887, carved with a knife blade by a man hiding in a ravine at the north edge of the Hills in 1834. On one side, in carefully incised letters: "Got all of the gold we could carry our ponys all got by the Indians I hav lost my gun and nothing to eat and Indians hunting me." On the other side, a list of names and this: "Came to these hills in 1833 seven of us all ded but me Ezra Kind Killed by Ind beyond the high hill got our gold in June 1834." Subsequent detective work suggests that two of the party may have come from the mining regions of Georgia and North Carolina, and that Ezra Kind had written his epitaph.

The Black Hills had remained mysterious for so many years at least in part because of their difficult geography. Punched up into an oval dome by upwelling magma, the Hills were a kind of hiccup of the Laramide Orogeny that had formed the Rocky Mountains. Erosion lopped off the top of the dome and cut deep canyons both concentric along the outer sedimentary rings, and radially from the crystalline core, a spider web of topography. General Custer did little more than skirt the edges of the northern Hills, turned back repeatedly by impassable canyons and thick timber. The prospectors of '75 had to weave their way carefully through the maze of difficult terrain, testing the gravel in this gulch and that stream until they finally hit pay dirt in December along a creek named Whitewood and a gulch called Deadwood for the burned and fallen timber along its course.

When twenty-year-old Richard Hughes slid down the steep hill into Deadwood in May of 1876, he found the creeks and gulches claimed from top to bottom, log cabins sprawled haphazardly, and men placer mining what ground wasn't inundated by the spring runoff.

Hughes had quit a school teaching job in eastern Nebraska, anxiously finishing his spring term as he dreamed of the golden Hills. "The winter had seemed a long one," Hughes wrote, "and my thoughts had been busy with the contemplated great adventure, rather than the duties of a pedagogue, so that I fear even the exceedingly meager salary paid a country school teacher at that time in Nebraska was fully commensurate with my services in the cause of education."

Slender, fine-boned, and frail looking, Hughes had worked as a printer's apprentice for three years before taking up teaching. The wagon boss in Sidney, where Hughes got off the train with two friends and asked to hitch a ride, said "if that young fellow ever reaches the Black Hills he'll not stay very long." And Hughes admitted "my four years of indoor occupation had not tended to give me a rugged appearance." The wagon boss took him along anyway, and they set off northward through the prairie. The route from Sidney passed through the Red Cloud Agency, where soldiers warned them of the conspicuous absence of most of the Sioux young men. The night watch was tense and wagon trains bunched together for protection.

All along the route, Hughes met returning stampeders, nearly two hundred fleeing the Hills. Like the busted of Pike's Peak, they had not found the streets of Custer paved in gold, but more ominous still, the many miners who prospected northward had never returned, and must, therefore, have been slaughtered by Indians. The natives indeed were causing havoc around the Hills, picking off stragglers and sending sniper fire into the very streets of Custer, which was practically abandoned when Hughes reached it fourteen days out from Sidney.

Custer, in the spring of 1876, was a town of five hundred houses strung through a natural park. "The population, which a short time previously had numbered more than a thousand souls, had dwindled to a few hundred, and it seemed that even a majority of those were intent on leaving," Hughes observed. His party moved on as well, heaving their wagons northwest over steep ridges and across boggy streams, on a barely traceable road filled with stumps and strewn with rocks. The final descent to Deadwood was so steep the wagons had to be lowered with guy lines wrapped around nearby trees. Hughes then thought he knew why the northern miners never returned. "While it was possible to get down [the hill] by means of ropes," he said, "no one would ever be foolish enough to climb [back up] it."

The other reason the miners stayed put was the fabulous wealth in the narrow little drainages, weathered out from surrounding hard rock lodes. The better claims were paying from $150 to $1,000 per day, bedrock was shallow, and water was plentiful. By June, 1876, a cumulative wagon load of gold worth $500,000 was shipped out to Cheyenne, and the year's total was estimated at $1.5 million. For men like J.J. Williams, who had been among the discovering party, times were grand. His Number Two Below Discovery claim turned out $27,000 before he sold it for $1,000. He then bought Number Fourteen Above Discovery and washed out $35,000 more.

For men like Richard Hughes, who arrived after all the good claims were taken, life was a struggle. He and his friends built a brush "house" with a bark roof along the crowded banks of Whitewood Creek, and set off prospecting in the surrounding hills. Because so many were likewise underemployed, stampede mania was the order of the day. Hughes had joined a partnership that agreed to send out one representative per rumor of rich new diggings while the rest stayed put, which demonstrated a certain level-headedness amidst the general madness. More often than not, Hughes turned out to be that representative. Almost immediately he was sent off to "Potato Gulch" thirty miles to the west. The gold there was thin and fine and not worth digging.

In June, Hughes dashed off to "False Bottom Gulch" where rumor said there was gold beneath a false bedrock. It was the rumor, not the bedrock, which was false. Later that month Hughes was lured to Polo Creek when "friends" informed him "in strictest confidence" of a rich new find across the mountain where he should meet them the next day. "I was directed to be at the mouth of Blacktail Gulch the next morning at daylight. There I would find the track of a sharp shod mule leading up the gulch a mile and a half, where it would turn to the right along a dim trail and over a low divide to the head of the gulch in which the discovery was made." When Hughes crossed the divide he saw to his astonishment, not the three original discoverers working a lonely prospect, but hundreds crowding around a tent, "while as far as the eye could see along the course of the gulch men were engaged in staking claims, posting notices and digging prospect holes." Inside the tent was the elected recorder collecting $1.50 for every claim. Whether said recorder was responsible for this hoax cannot be verified, but his name was David Tallent, husband of Annie Tallent, returned to the Hills to recoup his earlier losses. After questioning several of the stampeders, Hughes concluded that "the news of a discovery and its location had been imparted to many, as it had to me, as a secret, and a stampede [to goldless ground] was [thus] assured."

These local stampede hoaxes were relatively harmless compared to the one perpetrated the following winter that caused hundreds to rush off to the imaginary Wolf Mountains. "This stampede was a crime, and a damnable one," Hughes recounted, "growing out of the greed of a few individuals who had no regard or consideration for the misery and suffering it caused. That the rumors ... were set on foot by men who had, in advance, bought up a large number of ponies suitable for pack animals—expecting to sell them to the stampeders—has been pretty well determined." For once, Hughes was not the chosen representative, and the partner who did go promised to return in ten days if nothing definite developed.

"The fact was," Hughes reported, "that no one seemed to know whether the so-called Wolf Mountains were a part of the Bear Lodge, the Big Horn, the Wind River Mountains or the main range of the Rockies. It may seem inconceivable to a reader at this day, who never has witnessed the effect upon human beings of the stampede mania, that men—hundreds of them—yes, whole communities—could become so lost to reason as to start off, in the dead of winter, to an unknown but supposedly distant region, without knowledge of how it might be reached, and with no foundation for their faith in its richness stronger than the most vague rumor."

As promised, Hughes's partner returned in ten days, "firmly convinced that none of the stampeders possessed more knowledge than himself. Thus he escaped much of the hardship endured by many others, who persisted in going farther. Of those, some were killed by Indians.... A few reached the mining districts of Montana and wintered there.

One party spent the winter on the Crow Reservation. . . . A few arrived [back] in Deadwood, after having endured many dangers and hardships, and in a generally dilapidated condition."

Between stampedes, Hughes and his partners continued to prospect, locating a hard rock lode on the mountain across from Deadwood that turned out to be very valuable, but only after Hughes had to sell it for $100 to stake his partner's participation in the Wolf Mountain stampede. "In the life of the prospector there is constant hope," reflected Hughes, "and no matter how often he may have been disappointed he is never utterly discouraged."

Hughes turned his attention to their other discovery, a placer claim in a dry gulch feeding Deadwood Creek. One pan carried down to the creek would show good color, while the next would reveal only a few specks. To really work the claim it was necessary to dig a ditch from high up Deadwood Creek, build flumes across two intervening gulches, and run a sufficient head of water through their sluices. It was a gamble they weren't sure would pay off. As a final test they hauled several wagon loads of gravel down to the creek and washed them out in a rocker. "The result was, on the whole, encouraging," Hughes reported, "sufficiently so as to decide us in favor of the ditch." It was a bit like Russian Roulette, this game of gold hunting.

They spent an entire month hand digging the ditch along the course of the hillsides, scraping together enough money to buy lumber for the flumes, and when the time came to run water in it, Deadwood Creek had dropped enough that they could only get water at night when other miners weren't using it. The ditch immediately washed out and ruined an angry miner's garden below. The partners scraped together more money to pay damages, repaired the ditch, and finally were ready for their first run of sluicing. "We looked forward with mingled hope and fear to the time set for the first clean-up" Hughes reported. "I do not care to dwell upon our feelings when that clean-up was panned down and the result announced. It was so pitifully small that we realized at once that all our labor and expense, all our hopes and plans, based on belief in the value of the ground, were vain. In my lifetime I have met with disappointments of various kinds, but I can recall none that hurt as deeply as this."

Like another famous would-be miner, Hughes went to work for the local newspaper, and writing, not gold, would be his living. It must have been an especially hard decision, with so much sudden wealth surrounding him. "By the middle of June," Hughes recorded, "Deadwood, which had become the objective point of all travel into the Hills, presented a scene of great activity. The placers that had been opened on Deadwood, Whitewood, Gold Run, Bobtail and Blacktail Gulches were turning out large amounts of dust, which many of the miners were spending in the manner of the proverbial drunken sailor—in drinking, dancing, gambling and every form of carousel offered. Wages to shovelers on bedrock were seven dollars per day or night shift. Such pay to a lot of young fellows from the States—many of whom had never been paid more than fifty dollars per month—made them feel almost rich."

There were surer ways of winning gold than grubbing it from the ground, as the artists of purveyance who swelled the skirts of every gold rush had learned. They arrived early in Deadwood: "Of the structures fronting Main Street, the majority occupied by the last of May were saloons and gambling houses, where practically every form of gambling known in America was conducted. Such places never closed their doors, day or night," Hughes observed. Next arrived dance halls, then theaters. Mine shafts were sunk in the very streets of Deadwood, a town squeezed unnaturally into a tight-walled canyon and stacked up the hillsides until there were stairways instead of streets.

It was into this melee of swinging picks and loud-mouthed monte dealers, seven-yoke ox trains and a crowd of restless humanity that Bill and Jane rode one day in July of 1876. If ever two people had seen it and done it, they were the ones. Calamity Jane had become infamous for her hard drinking and a mannish tendency to wear buckskin trousers and drive bull trains, not to mention a certain immodesty where her virtue was concerned. "Wild Bill" Hickok had a deadlier reputation. His career as scout and sniper for the Union Army, stage driver, law man, and friend of General Custer and Buffalo Bill Cody had earned him the sobriquet "Prince of Pistoleers," as well as a starring role in numerous magazine articles and dime novels that exaggerated his exploits beyond the limits of credulity.

But the man who rode into Deadwood that July was moody and apprehensive, thirty-nine years old, his eyesight failing. He was hoping to catch one more glimpse of the wild, wide open frontier he had thrived on, a frontier closing with the century and locking its gates. Though recently married, Hickok arrived in Deadwood *sans* bride, riding the length of Main Street on a fine horse and wearing a buckskin suit with "sufficient fringe" as Richard Hughes described it, "to make a considerable buckskin rope." He wore his hair and moustache long and flowing to frame his deep-set eyes, chiseled cheek bones, and gentle, rounded chin.

By his own count, Hickok had killed thirty-six men in fifteen years, managing, just barely, to stay on the safe side of the law. Jack McCall had killed no one and done nothing in his lifetime worth mentioning, and he would pay the ultimate price for this, his one streaking comet of fame. Hickok did not know McCall, but McCall knew Hickok. Richard Hughes gave this report of events. On arriving in Deadwood, Hickok "naturally fell into the practice of his profession as a poker player." He was playing in Nuttal & Mann's Number Ten Saloon on August 2nd, "sitting with his back to the door when McCall, who had been in the camp for some time but was little known, entered, stepped quickly behind Hickok, and with a revolver of large caliber shot him through the head, killing him instantly." Hickok slumped over on his cards, black aces and eights, known ever after as "the dead man's hand."

McCall was tried in Deadwood by miners' jury, and acquitted by the usual rabble that wasn't quite able to connect crime with punishment. He left town but was tried again in Yankton (on the grounds that the miners' jury was not legally binding), convicted, and hanged.

Hickok's grave now rests in Mt. Moriah Cemetery, high on the steep ridge where the view of Deadwood, false-fronted and patchwork-backed, is foreshortened in time as well as space. Ponderosa pines shade the ground, and tourist trails weave in and out of the pale white headstones. I looked there for a moral to this story beyond the prosaic—that men of his class generally die with their boots on; he played out his hand; it was one more wicked deed in a wicked town. But Hickok's death had nothing to do with Deadwood. Gold rush murders were crimes of passion—anger, fear, greed, revenge. Hickok was a victim of his own notoriety, a tabloid-generated target, just one in a trend.

As a boom town, Deadwood had pretty much run its course when five-year-old Estelline Bennett rode the stagecoach into town. Burned to the ground in 1879 and risen, brick by brick, from the ashes, Deadwood had run out early of its rich placer gold, and the town was living off numerous hard rock mines (which also would dwindle or be consolidated into the one great mountain of gold called the Homestake). But enough of the rawness remained to keep Estelline, the daughter of a federal judge, preoccupied with the fine differences between the seamy and the respectable. "Deadwood was a wide-open town in the eighties with saloons and gambling houses and painted ladies—its spectacular vices on parade for all the world to see," she explained.

While Deadwood had its respectable neighborhoods tucked into the hillsides and "sheltered as the courtyard of a convent," down below within view through the narrow alleys lay the "badlands" and Chinatown, with all the drunks and bummers and unregenerated dregs of a used-up rush. The stampede to the Black Hills had been small by California or Colorado or even Montana standards, with a peak population of fifteen to twenty thousand, and a production of $12 million during the first five years of the boom. But in Estelline's eyes, "of all the mining camps of the frontier, Deadwood flared highest and brightest." Manic in the crush of a closing frontier, condensed in space and overwrought from three decades of rehearsal, Deadwood would indeed remain vivid in the American imagination and well lit in the mind of a girl who observed and meditated upon its stock cast of characters.

There was Swill Barrel Jimmy, "his long frock coat turning all the sad apologetic shades of purple and blue and green when it should have been black," who lived out of trash cans in the alleys, spoke in monosyllables if at all, and spent his few pennies on clean, white paper collars for his worn out suit.

And the girls of the badlands, "lovely light ladies—pretty, beautifully gowned, and demure mannered ... [who] collected the wages of sin under our very eyes. We saw them on the street, in the stores, at the theater on the rare occasions when there was a play at Nye's Opera House or Keimer Hall. And then, two or three years later, we saw these same girls pallid and shabby ... slipping furtively down the alley that ran back of Main Street."

There were gamblers, "the lily-fingered leisure class of Deadwood ... who seemed never to toil nor spin, yet always were arrayed in fine linen and broadcloth fresh from the tailor's iron." Gambling and drinking were the favorite pastimes in gold rush towns, but all such carousing was momentarily suspended one evening in Bedrock Tom's saloon as a bare-headed woman strode up to the table where her husband was losing at poker, and swept the large center stake of gold coins into her gingham apron. "In the deeper silence that followed [her bold act], while not a card fluttered, not a foot moved on the floor, and the roulette wheel and the clink of glasses were stopped, she strode out with her booty, and they let her go as though she had won it in a square game. No one ever mentioned it later or questioned her right to it. But never again did any gambler in Deadwood Gulch play poker with *that* woman's husband."

As Estelline explained, "law and order came early in '77 and for more than a dozen years lawlessness and law, order and disorder drifted along together in the seclusion of the deep, narrow gulch without interfering seriously with each other. They emphasized the natural sharp contrasts of the camp in Deadwood Gulch." The courts were tied up with the intricacies of mining law, complicated by a maze of underground maneuverings where shafts and adits crossed property lines in the dark.

Above ground, more than thirty murders were committed in the first three years of the boom (Hickok's was a notable third in a long string), while over a hundred people (mostly coming or going) were killed by Indians. As the Indian threat lessened, robbery overtook murder as the favorite crime, with road agents waylaying the stagecoaches that provided Deadwood with its only link to the outside world. Stagecoach holdups became so common that passengers expected to be robbed and left their valuables at home. Gold and other precious cargo was shipped in a special treasure coach, "a steel-lined, iron clad, port holed fort on wheels." The treasure coach was successfully robbed only once, and most of the stolen gold was eventually recovered.

Holdups were a thing of the past when Estelline Bennett began riding the stage in and out of Deadwood, the perpetrators hanged or shot or sent to prison. But the idea of danger was still fresh in her mind, incited by the stories of dashing drivers who helped her imagination work overtime. "As each stage station was slowly sketched out of the brown plains by the narrowing distance, I expected to find the long low cabins deserted and adorned with rows of bloody scalps while red men in war paint danced, and white men in black masks waited behind the log walls to spring out and add us to their bloody clutter of silent victims."

But Estelline admitted "I had the personnel of the enemy darkly confused in those early days. Indians and road agents were alike to me—wild men who lived only to devour stagecoaches and their occupants." As Estelline grew older she began to make a distinction between the two bands of marauders, between the outcast white men who cowardly chose crime over a more acceptable profession, and "the robbed and resentful wards of the nation, one-time owners of all the West."

The Sioux who had been one-time owners lost the Black Hills in their last and most valiant fight, and it was gold that tipped the balance against them. The government would never have launched its massive military campaign in 1876 if gold hadn't been found on Sioux land. When bribery failed to secure the Hills, the army decided to beat the Sioux into submission. The campaign designed by General Sheridan would bring forces together from three sides—General Alfred Terry's column from the east, Colonel John Gibbon's from the west, and General George Crook's from the south—to crush the Sioux's summer camp at the north end of the Powder River basin. (The campaign was supposed to begin in winter, but the army had enough problems without battling bad weather.)

Crook was the first to engage the Sioux in a mutual surprise attack on the headwaters of the Rosebud (scouts from both sides actually ran into each other on a ridgetop and went scurrying back down to warn of attack). With heated fighting between fairly equal numbers, both sides eventually retreated with surprisingly low casualties. General Crook was declared the official loser because he retreated farthest and longest, effectively withdrawing from the campaign.

Terry and Gibbon managed to meet up on the Yellowstone River, and gathered intelligence that the Sioux had probably moved from the Rosebud to the Little Big Horn. Gibbon was to hold the northern line at the Yellowstone, Terry would march up the Big Horn, and the 7th Cavalry under Bvt. General George Armstrong Custer would sidle up the Rosebud to cross and descend on the Sioux from the south.

Looking through a spy glass from the Crow's Nest on the divide between the Rosebud and Little Big Horn drainages, Custer couldn't quite make out the Sioux village tucked into the beautiful valley of the river below, but his Crow scouts said it was there, plain as day, and bigger than any village they had seen. Custer's troops had made a forced night march up the divide and were resting in the timber below, but it appeared that they had been spotted, and would have to attack immediately to maintain any element of surprise. Over the ridge they poured, 597 officers and enlisted men with 50 additional scouts, packers and civilians.

Custer began his attack by splitting his forces four ways, taking five companies himself, giving three to Major Reno, three to Captain Benteen for a reconnaissance to the south and west, and leaving one company with the slower moving pack train. Near the south end of the village, Reno was ordered to cross the river and charge, while Custer continued along the eastern bluffs to attack the village in the middle. Reno was met immediately by a defensive swarm of Sioux and his ordered retreat turned into a rout as his troops fell back to skirmish in the timber, then fled across the river and scrambled up a steep ridge to regroup at a better firing position. Less than three miles away, Custer divided his forces again, sending two companies down Medicine Tail Coulee in a feint at the center of the village, while he continued north to surround the village at its head. This move drew the Sioux away from Reno, and after Custer in full force.

Custer had sent two terse messages back to Benteen to bring reinforcements and the ammunition pack mules, but Benteen, on reaching Reno's hilltop and hearing heavy firing in Custer's vicinity, did not want to venture further. All of the Indian scouts, except a few who had been trapped or killed, had grabbed some loose Sioux ponies and gotten the hell out of there.

Disgusted with the timid responses of Reno and Benteen, Captain Thomas Weir rallied the soldiers on the hilltop and advanced a mile ahead, intending to relieve Custer. But the Indians had finished with

Custer and were turning back toward their other quarry. Weir and company retreated back to Reno's hill and all held out for a day and a half until the Indians packed up and moved away from Terry's slowly approaching column. Fifty-three of Reno's and Benteen's soldiers were killed, and 210 with Custer, including two of his brothers, a nephew and his brother-in-law.

Representations of Custer's Last Stand are largely erroneous in showing hundreds of soldiers overrun by a mob of Indians and fighting valiantly to the last man. The Sioux were smart and cautious warriors, not prone to suicidal charges when the odds were in their favor, and never had the odds been so much in their favor. Recent archeology suggests the more likely (though less picturesque) reality is that Custer's command was individually picked off at various defense positions, too spread out to back each other up, until only a bare company, rallied round Custer, remained to be fired upon.

After they left the Little Big Horn, the wild Sioux were harassed the rest of the summer by a worn and dogged army propped up by an outraged nation. The reservation Sioux were threatened with starvation if they didn't sign away the Black Hills. The vise had finally closed completely, and the only apparent justice was that the man who had started this last war, the man who thought he would finish it, had finished himself.

When historians piece together the Battle of the Little Big Horn, laying one account beside another, one sliver of evidence and slice of contradiction, what they wish for most is a view into the mind of Custer on that day. What could have motivated him to act as he did? We can glean certain assumptions he may have made. Custer knew for certain there were at least eight hundred warriors in the camp and should have suspected as many as fifteen hundred, which means he would be outnumbered two or three to one. (Most estimates list two thousand warriors at the village, though not all of them participated in the fight.)

Both Custer and his superior officers were operating on the assumption (or apprehension) that the Sioux would run away, and the army would have to admit that most humiliating kind of defeat, that the enemy had slipped through their fingers like sand. Every tactical move was predicated on the fear that the Indians would escape, and on the conceit that the army could whip the Sioux if they would only stand and fight.

Custer's orders were ambiguous enough to interpret in a number of ways—to attack, or simply to observe and wait for reinforcements. "The Department Commander," wrote General Terry, "places too much confidence in your zeal, energy and ability to wish to impose on you precise orders which might hamper your action." Their confidence must have been based on Custer's Civil War record, because his only real Indian fight was a winter morning attack on a peaceful Cheyenne village camped on the Washita, where women and children were slaughtered along with the old chief, poor Black Kettle who had survived Sand Creek. Custer, like Fetterman and other Civil War veterans, had little respect for the Indians' fighting ability, until it was too late.

But what if Custer had succeeded? A rising star catapulted by one grand battle, a flawless military deployment against overwhelming odds, the man who beat Crazy Horse and leveled Sitting Bull? Ambition can take a person a long way down the road, but it can also make him miss the signposts.

Chapter 11

A YEAR AFTER the demise of Custer, prospector Ed Schieffelin looked carefully across the broad San Pedro valley, and you might have seen the hair stand up on the back of his neck, if Ed's tangled black locks didn't hang to his shoulders. He glassed the cut of hills rippling the horizon between the southern Mule Mountains and the northern Dragoons, sweeping nervously to left and right for a flicker of movement or puff of dust that meant Indians. It was not the grass prairie with scattered sage on the ridgetops that was Custer's last vista, but Sonoran desert plants—mesquite and ocotillo, saguaro and palo verde—deceptively green in the wet season and crisped to burnt umber in the dry. And it wasn't recalcitrant Sioux Ed was looking for, but the uncontrollable Apaches. Led by the likes of Cochise and Geronimo, the Apaches walked on and off their reservations whenever the mood struck them, raiding across Arizona and into Mexico.

Ed could see nothing amiss out there where the San Pedro River was only a suggestion, its water hidden in the valley's central crease. He cared more about the rocks at the edge of his view, a rumble of hilltops in the realm of the possible, than he cared about the Indian threat. "I wasn't looking for bullets," he would comment later about his bold invasion of the Chiricahua Apache stronghold, "but I felt if one happened my way it wouldn't make much difference to anyone but me." The soldiers at nearby Fort Huachuca thought a man ought to have more concern for his own hide. They told Ed that the only thing he would find in those southeastern Arizona hills was his tombstone. "I never could figure out that to be dead would be unpleasant," Ed concluded. He was unusually pragmatic for a man of vision.

Ed Schieffelin had not yet turned thirty when he reached this arid and dangerous corner of Arizona, though he looked at least forty—his matted beard as long as his hair, and his clothing so patched with deerskin and flannel that there was little of the original material left. He was just a boy when he left Pennsylvania with his father for the Rogue River mines in Oregon, and he had been a prospector ever since, wandering from Nevada to Idaho, California to Oregon and back again, always looking for the main chance and stopping just long enough to gather a grubstake before moving on.

Arizona was the one place Ed hadn't searched, and when he reached the San Pedro valley, it was as if a star shone down on the line of hills at the eastern edge and only he could see the light. He ventured farther and farther into those hills and away from safety during the summer of 1877, collecting chunks of ore (called float) that erosion had separated from ledges of treasure concealed in the hillsides. And like he had some divine protection, the Apaches never bothered him. At the end of the summer he staked out a claim he satirically named "Tombstone," and set off to the north to find another grubstake.

Ed's brother, Al Schieffelin, was working at a mine near Signal, Arizona, when Ed showed up there the end of September. While Ed was a prospector, Al was a miner, and thus was not prone to flights of imagination, and was certainly not willing to give up a paying job if he could help it. But Ed chipped away at him, still seeing the light on the otherwise drab hills. He showed Al his ore samples, and finally convinced him to pay for assays. A highly-regarded mining man named Dick Gird had signed on as assayer for the Signal mine, and when Al brought the samples to Gird for testing, they ran $2,000 to the ton. "The best thing you can do," said Gird to Al Schieffelin, "is to find out where that ore came from, and to take me with you."

The three men left on Valentine's Day, 1878, traveling four hundred miles across Arizona by wagon. By the end of the summer they had staked out half a dozen claims with names like the Lucky Cuss, Tough Nut, and Contention. It wasn't gold Ed had found, but silver so rich and pure that in one outcropping he pressed a half dollar into the vein and when he pulled it out the mirror image was as crisp as the original.

Silver had increasingly become the metal of boomtimes in the closing decades of the nineteenth century. Gold was easier to find by far, advertising itself at the bottom of streambeds and leaving a trail like bread crumbs back to its source, offering instant reward to the placer miner and simple separation to the hard rock miller. But silver wore a more elusive robe and attended her subjects with less immediate finery. A person needed money to make money from silver ore, and if this were the usual story of prospectors, Ed and Al would sell out for several thousand dollars and drink up the profits. They did sell one of their claims, but used the proceeds to develop the others.

It was a long and involved ordeal, opening a silver mine. They needed hundreds of thousands of dollars, and for that they turned to Eastern financiers who formed corporations and sold stock. Dick Gird went to San Francisco to purchase a stamp mill, while Ed and Al Schieffelin hired miners to begin working the silver ledges. The stamp mill was freighted in pieces by wagon, and required a dam across the San Pedro River as well as a good road up to the mines. The first bullion brick, from ore pulverized in the stamp mill to a slurry, mixed with mercury, settled, retorted, then purified in a smelter, was finally poured in June of 1879.

Ed himself accompanied the first weekly shipment of bullion to a bank in Tucson, hauling it in the wagon they had driven from Signal, and hiring three men to guard the $18,000 treasure. But Ed was getting tired of this mining business, and in November he loaded up his mules and left the booming town of Tombstone in search of another prospect in the Arizona hills.

Platted on a mesa top called Goose Flats, Tombstone would flare and burn out in a decade, but it was lively while it lived. Fourteen working mines turned out $500,000 monthly, and three thousand claims staked across the forty-square-mile district encouraged the dreams of its ten thousand residents. But in the end, it wasn't the $25 million in silver bullion that Tombstone became famous for. It was its lack of lawful order, fueled by remoteness, fired by incendiary personalities, and ignited in the infamous "Shootout at the O.K. Corral."

In the Hollywood version it became a simple story of good against evil, the Earp clan facing off against the rustling Clanton and McLaury gang to win the day for the civilized world. Wyatt Earp was a brave and virtuous man, his brothers upstanding citizens, and his friend, Doc Holliday, fearless if nothing else. The "Cowboys," represented by the Clanton and McLaury brothers, threatened to take over the whole town of Tombstone in *High Noon* fashion and only the Earps stood in their way. Or so we would like to believe.

What actually happened was gritty, unpleasant and in no way clear as to who were the good guys and who the bad. The shootout took place on October 26, 1881, in a vacant lot half a block away from the O.K. Corral (a kind of horse parking lot), and when it was over three bodies were displayed in elaborate coffins with a sign in the undertaker's window reading "MURDERED IN THE STREETS OF TOMBSTONE." Wyatt Earp, older brother Virgil Earp, the younger Morgan Earp, and John Henry Holliday were put on trial for murder, though Virgil and Morgan could not attend because they were suffering from gunshot wounds.

The trial unfolded in a maze of contradictory statements from participants and eyewitnesses. Everyone agreed that the churlish Ike Clanton had gotten into an argument with Doc Holliday and Morgan Earp in a saloon the night before. Ike and brother Billy Clanton were frequent participants in the sport of raiding across the border for Mexican cattle; their father had been killed two months earlier in a Mexican counter-raid. Ranchers Frank and Tom Mc Laury were associates of the Clantons, but their guilt as rustlers remains uncertain.

The Earps had a checkered background as well. Wyatt had been a lawman in the cattle boomtowns of Wichita and Dodge, but he'd also fled a horse-stealing charge in Arkansas and had been kicked out of Wichita for "vagrancy." Virgil arrived in Tombstone with a deputy United States marshal commission, which he used to further deputize his brothers whenever the need arose. Doc Holliday, a trained dentist with the thin frame and incessant cough of tuberculosis, had lost his fear of death and liked to make enemies. Morgan Earp, the handsome younger brother, liked to hang around with Holliday where life was never dull. All of the Earp clan made their living in Tombstone by gambling, a profession which oddly seemed to go hand in glove with being a lawman back then.

Ike Clanton had been fuming death threats against the whole Earp clan on the morning of October 26, and the whole Earp clan marched out to meet him. The Clantons and McLaurys were trying to leave town when the showdown came, and only young Billy Clanton and Frank McLaury were armed. Ike Clanton ran like a dog for cover when the shooting began; Billy and Frank put up a good fight, but they and weaponless Tom McLaury were riddled with bullets in less than thirty seconds. Virgil Earp was shot through the leg, and Morgan the shoulder, but both would recover.

In the trial that followed, the Earps were acquitted of murder, but the cowboy friends of the Clantons and McLaurys sought their own justice. Virgil Earp was ambushed in downtown Tombstone on December 28, his left arm shattered by buckshot. The following March, Morgan Earp was shot dead while playing billiards in Robert Hatch's saloon, the bullet smashing through a glass-windowed back door. Wyatt would shortly thereafter gun down ex-deputy Frank Stillwell in the Tucson train station. The violence continued until what remained of the Earp clan finally had to flee Arizona with a posse on their trail.

No true heroes and no true villains lived in Tombstone, it seems. It was a town where the newspaper was wryly named the *Epitaph*, and whose Boot Hill was crowded with the markers of people who took a wrong turn. The grave marker of Ed Schieffelin, who started it all, rests not in Boot Hill but on a lonely hillside near his original discovery. And the marker is not a tombstone, but a piled rock "monument such as prospectors build when locating a mining claim."

Ed had returned to Tombstone in 1880 with the realization that there was a difference between looking and finding. He had moved, in the Tombstone hills, from the realm of the possible to the land of the probable, with its valleys of doubt, and he didn't like it there. So Ed sold out his share of the mines for $300,000—one giant grubstake for the rest of his life. He loaded his mules and wandered away toward another shining star, dying of a heart attack in Oregon at the age of forty-eight with his last request to be buried near Tombstone.

"I am getting restless here in Oregon," Ed had written to a friend shortly before his death, "and wish to go somewhere that has wealth for the digging of it. ... I like the excitement of being right up against the earth, trying to coax her gold away and scatter it." Trying to coax her gold away and scatter it, the words keep echoing in my mind, coax her gold away and scatter it.

THE END OF JULY was unbearably hot here, sky cloudless and sun burning through windows and the uninsulated roof of my cabin. There was no relief outdoors, for the willow bushes occupied their own shade and the aspen offered only trembly shadows. I opened both doors for a whiff of breeze (where is the wind here when you need it?), and a rock wren flew in one day and perched on the ceiling cross beam. She examined my furniture—a table with two rickety chairs, a bed, a Naugahyde recliner propped against the wall to keep it from reclining permanently, a refrigerator that didn't refrigerate, a sink that wouldn't run water, two wood stoves gone cold, and me. She was polite enough to keep her opinion to herself as she flew back out the door.

To escape the heat that was slowly draining color from the parched landscape, I headed the first of August for the mountains between the Wyoming towns of Pinedale and Jackson Hole. High piles of rubble, these mountains are steep, timber-clothed ridges whose drainages run unpredictable directions, the place where Robert Stuart and the returning Astorians nearly starved to death in 1812. The mountains are a mixture of dirt, sandstone ledges and quartzite cobbles, the outwash remains of older mountains long vanished that were, a hundred million years ago, as rich as Croesus in pure gold.

I can dip my pan in any one of the creeks draining these rubble ridges and swirl out a tail of fine flour gold. The trick isn't finding gold, it's finding enough gold. The farther gold travels from its original source, the more it gets worked over and the finer it becomes. The gold here has been reworked two or three times—freed from the original mountains, tumbled along vanished streams, pounded in beach sand, raised into mountains again and washed into new streams.

It is my bright idea to short-circuit one of the washings by tracing a certain stream to its source on a high peak still holding its original deposition layers. Ninety years ago a Crow Indian claimed there was a "fabulously rich" mine half way up the 10,840-foot mountain, but no one believed him. Like most paleoplacers, it seemed an unlikely place to look for gold. I parked my pickup at the end of the road, crawled into the harness of my overloaded backpack, and started up a dim trail along the streambed.

It was primetime for the wildflower show along the marsh and willow choked creek, and the Indian paintbrush, which range in color from yellow through fluorescent orange to almost black, were here the purest shade of red I had ever seen. An abundance of moisture, lack of roads, and four thousand square miles of essential wilderness stretching up to Yellowstone Park also makes this prime bear habitat. My revolver was strapped to my waist belt, and I peered nervously ahead, startled occasionally by my dog leaping out of the brush.

I've had more than a dozen bear encounters in my lifetime outdoors, usually resulting in a mutual retreat at high speed. I once chased a black bear out of a cook tent with a butcher knife, though. He had sidled in the back door of the canvas tent while I was looking the other way, and when I turned around I found him hungrily contemplating a dozen lunches I had laid out on a table. It was a split second decision, one I might have altered given careful thought, but I charged instinctively and the bear, confronted by a screaming, flailing, knife-wielding lunatic, made a hasty escape out the back door and up the nearest tree where he draped himself on a branch, whining and snapping his teeth until I let him down half an hour later.

I try hard, as personal policy, to stay out of grizzly country, but I do not always succeed. Working as a wrangler near the Tetons I once watched a young male grizzly digging for pocket gophers in a hillside across a small valley, which was as close as I ever wanted to be. But a week later I went to pack out a bull elk a hunter had shot the day before, leaving my saddle horse on the ridgetop and leading two pack horses along a steep slope to where the elk was cached. I smelled before I saw that a grizzly had been there, a sour, rotten smell like unwashed clothes and old meat. The bear had found the pile of elk guts, rolled in it, gorged on it then drug it downhill and buried it shallowly so he could come back for more.

Time stopped and I could feel the hair raise on the back of my neck. My hand reached down for my revolver, pulling it from the holster and half raising it as I turned 360 degrees on my heels looking for the attack. I felt, at that moment, like a Clanton with the Earp boys coming to get me, and it would all be over in thirty seconds or less.

It is anticlimactic to tell you that the bear must have been sleeping off his night's orgy, but he never showed up. I made off with the rest of the elk, keeping a worried eye on my back trail, and I have since tried harder to stay out of grizzly country.

I was thinking these uneasy thoughts about bears while trudging up the creek valley when I stopped at the edge of a meadow to hitch my pack up and rest my shoulders. In the timber across the meadow I heard my dog yelp, and then there was crashing in the underbrush. The dog popped out of the timber running like his tail was on fire. Something big was coming after him, crashing and snorting. The dog saw me and changed course straight towards me, like I was going to save him. Out of the timber in full pursuit charged a cow moose.

"My God," I thought, "the only thing bad as a bear is a mad moose." I loosened my hip belt so I could drop the pack and started running back down the trail looking for a tree to climb. The moose finally saw me and stopped, suddenly wary. I stopped too, hoping bravado would force her back. The trail went right under her feet, and I didn't want to go around through the swamp. It seemed to be a stalemate so I made a feint at her, yelling and charging. She called my bluff, charging back. I retreated. Reinforcing myself, I charged again waving a stick. Unimpressed, she lowered her head and charged back. I retreated. The dog watched happily from the sidelines, as if this were the grandest game he'd ever seen, tongue lolling and eyes sparkling. I tried a subtler approach, sauntering forward and studying the wildflowers like I was an innocent bystander. The moose stamped her foot and snorted; she wasn't born yesterday. I sauntered quickly back into the timber. By now I was beginning to doubt my poker playing ability. I pondered the stakes, studied my hand, and folded. The moose had a smug look on her long face as I set off through the swamp going the long way round.

It was almost dark when I reached the last patch of trees on the mountainside and pitched my tent on the alpine slope. The creek here was just a trickle of meltwater, and in the morning I followed it up through patches of snow and the fresh mud of newly bared ground. My aim was a cut bank at the top of the ridge, its layers of sediment exposed in a ten-foot wall of dirt. Struggling up the steep gully, I found the vertical wall precariously unstable, but I managed to climb high enough to scrape a layer of dirt and cobbles into my gold pan. I trudged back down the slope far enough to dig a pool in the meltwater and wash out the dirt, and there in the bottom of the pan was gold.

But was it enough gold? I looked back at the high wall, looked around at the infinite number of similar summits any of which could contain an old streambed with rich, undiscovered pay streaks of gold, as if a giant Midas had touched them all. It could take a lifetime of looking, and what if, in that bear and moose infested wilderness, I actually found something? The idea was unnerving enough to make me pack up the next day, head down, and drive back to the desert.

PERHAPS IT WAS inevitable, after centuries of telling, that the moral of the story of King Midas has been forgotten. The original fable was a cautionary tale against greed, but we remember only that Midas had a golden touch, as if that were a blessing instead of a curse. There are more contemporary fables that remind us of the mutable nature of fortune (and fortunes), like the story of the cowboy and the carpenter at Cripple Creek, Colorado. Both were cursed with different versions of the Midas touch: one couldn't get gold to stick to him, and the other couldn't get unstuck.

Bob Womack was a cowboy, but not a very good one. Oh, he could ride, and he could punch cows, and he was better than many at spending months at a time alone on the open range, but most of the time most of his mind was occupied with something other than cows. It was just as well, for if Bob Womack had simply been a cowboy, no one would remember him. But Womack was also a prospector, trained in the Pike's Peak Rush and unable thereafter to pass a rock without looking at it. His family had made a minor fortune at mining in the 1860s, and settled down after the rush on a ranch not far from Colorado Springs. Their summer cow pasture was a ten-thousand-foot-high bowl in the mountains with a stream running through called Cripple Creek. Bob Womack spent thirteen years picking up rocks around Cripple Creek that didn't look like gold ore; in fact they looked curiously like chunks of drab gray concrete, but they assayed (when the chemist didn't derisively throw them in the trash) a respectable $200 to the ton.

Nobody believed Crazy Bob had found a gold mine. It was hard to take the gangly cowboy seriously. He drank too much and talked too loud and couldn't explain why Cripple Creek didn't look like any other mining ground in Colorado. The rich placer trails that led like bread crumbs up to most hard rock mines were missing at Cripple Creek, a misleading fact if you didn't know that the high-altitude, grass-covered bowl was an old volcanic blowout, eroding in on itself and riddled with veins of the unusual gold ore called sylvanite. But Bob Womack believed. Year after year he followed the float—small pieces of gray ore—uphill until they disappeared. Then he dug down into ten feet of topsoil and rubble searching for the source vein and one day in 1890 he found what he was looking for.

The residents of nearby Colorado Springs—a resort town established in 1871 for the rich, the bored, and the infirm—still didn't believe in Bob's mine, at least those fashionably in the know. So the small rush to Cripple Creek in the spring of 1891 was made up of the unfashionable: a plumber, a grocer, a handyman, a furniture salesman, two druggists, a butcher, a roustabout, a lumberman, a schoolteacher, and a carpenter. The carpenter was also a prospector who had wasted fifteen years of his life searching through Colorado's mountains for a paying mine. He would have been fed up with prospecting if it weren't for the solitude, the way his burro listened and didn't talk back, the possibilities that rode along every day with the search. He had spent $20,000 of his own hard-earned money, carpentering in the winters to support his summer prospecting habit. When he arrived at Cripple Creek he was looking for cryolite (a more prosaic mineral used to make aluminum), not gold. He had pretty much given up on gold.

But Winfield Scott Stratton stopped by to visit Bob Womack at his shack in Poverty Gulch, and Womack implored the carpenter to stake a claim nearby. He liked the company, and was enjoying his sudden notoriety as discoverer of Colorado's newest Eldorado, never mind the fact that no gold to speak of had yet come out of the ground. Stratton staked several claims as far from the growing crowd of tenderfeet as he could get, then tried to sell them for $500, no takers.

So Stratton pecked away at his mine shafts, opening drifts and wearing a path to the assayer's tent. Other claims were showing good results;

the druggist had thrown his hat in the air, located a claim where it landed, and dug out a vein assaying an astonishing $600 a ton. But Stratton's ore was barely paying costs, so in June of 1893 he optioned his Independence mine for $5,000. If, after a month, the mine proved productive, the new owners would pay him an additional $150,000. Otherwise they would give the mine back. The next day Stratton was clearing his tools from the shaft, and in a drift abandoned the year before, he pulled down a drill rod and knocked loose a rock revealing a hidden vein. Stratton measured the vein and estimated three million dollars of ore in sight. But the contract was signed, and all Stratton could do was back out of the drift and hope the new owners didn't look there. It was a long month.

They hammered away at his other drifts to the same end Stratton had realized. The ore barely paid costs. The day before the option was up, they were going to start on the abandoned drift, but the manager wanted to leave town the next day and asked if he could give back the option one day early. Stratton couldn't keep his hand from shaking as they sat in the Palace Hotel in the new city of Cripple Creek. "Just throw it in the fire," he said. By December of 1893, carpenter Winfield Scott Stratton was Cripple Creek's first millionaire.

Twenty-seven more millionaires would emerge from Cripple Creek mines, many of them the original roustabouts and store clerks, lumbermen and butchers, handymen and grocers who believed in Bob Womack when no one else did. The population of Cripple Creek swelled from fifteen in 1891 to fifty thousand in 1900. Most of them were laborers in the 475 working mines, and those who labored for the laborers in saloons and stores and red-lit Myers Avenue. The problems that plagued Cripple Creek were of time, not place.

Unlike most gold strikes that had plunged into isolated frontier with its endemic lawlessness, Cripple Creek and nearby Colorado Springs were served by rail lines and all the latest amenities, including electric arc lights along the airy streets. The violence and anarchy that plagued Cripple Creek were the result of labor disputes common across America at the end of the nineteenth century, only magnified here by thin air and 625 tons of gold.

Winfield Scott Stratton's problems were of a different sort. He had sold his Independence mine in 1899 for $10 million, and he was a very unhappy man. Historian Marshall Sprague described Stratton when he first showed up at Cripple Creek in 1891 as "forty-two years old, a thin, pale fellow with silky white hair who looked as though he never got enough to eat." By 1899 he was fifty-one, thinner, whiter, and trying hard to get rid of his money. He gave a $50,000 Christmas bonus to each of his key employees. He bought and donated land for civic buildings and public parks in Colorado Springs where he had moved into a modest home he had built himself for a doctor years before. He spent $10,000 a day on worthless claims at Cripple Creek. He even handed $5,000 to Bob Womack because he saw him on the street. When Colorado Springs staged an elaborate banquet for philanthropist Stratton, 160 people showed up, but Stratton did not.

Like Midas with a golden drumstick stuck in his throat, Stratton felt like he was choking on his own fortune. A young man wrote Stratton for advice on whether to sell his claim or hold on to it for more money and Stratton advised him to sell. "I once gave an option on the Independence," he wrote, "and a thousand times I have wished that the holders had taken it up. Too much money is not good for any man. I have too much and it is not good for me." When Stratton died in 1902 at age fifty-four, he left $6 million of his estate to establish and maintain an elaborate home for poor children and old people. It was to be named for his father, Myron Stratton, a man Winfield Scott had run away from thirty-five years earlier to seek his fortune.

The Stratton Home might have had an ideal inmate in Bob Womack, the cowboy who started it all. If Womack had given up on Cripple Creek, the world might have missed out on the century's biggest gold strike. It was Cripple Creek gold, nearly five hundred million dollars worth, that tipped the balance against William Jennings Bryan and free silver coinage, and Cripple Creek ore that propelled America onto the gold standard. Bob Womack had faith in the gold, and such faith ought to be rewarded.

But Bob sold out his share in the El Paso discovery mine for $300. It was enough to set him up comfortably at his shack in Poverty Gulch, enough to watch the world grow suddenly big all around him and he was the sun shining out from its center. But those looking to the future are quick to forget the past, and Bob became one more bum with a whiskey bottle and a tall tale no one wanted him to tell. In 1893 he sold an old placer claim for $500, and as it was Christmas time, he had a few drinks and began handing out one dollar bills to every child on the street. "By degrees the children got taller," as Marshall Sprague described it, "and Bob saw that grown men were accepting his bills and were rejoining the line for more. Bob threw his fist at the very next face. Someone struck back and Bob fell to the sidewalk." Bob's sister, Lyda Womack, came to Cripple Creek and took Bob away to Colorado Springs where he became a cook in her boarding house, and died penniless in 1909 at age sixty-six.

In the tale of Midas, the king was instructed to wash in the Pactolus River to rid himself of his golden curse. The water cleansed his gifted hands and turned its own sands golden. If gold washed as easily from Bob Womack's hands, at least he had seen the golden sands.

CRIPPLE CREEK brought one of the few bright lights to the last decade of the nineteenth century, for the Gay Nineties weren't really so gay. A financial panic in 1893 had sent silver prices tumbling, closed mines, caused labor strikes and imposed a general depression across the country. Most people believed that Cripple Creek was the last great gold strike America would know; only the mountain's difficult geology had postponed its discovery so late in the century. And there were other things to be depressed about that decade. The U.S. Census Bureau declared the American frontier officially closed—they could no longer draw a definite line between the settled and unsettled. For a people who had thrived on the possibility of a better life beyond the known horizon, there was nowhere left to turn now, but back on themselves.

So the news was that much more stunning, in mid July of 1897, when two steamers arrived from the cold, gray Yukon River with three tons of gold on board. The men who staggered off the boats in Seattle and San Francisco, with their slouch hats and sunburned faces, muddy boots and ragged clothing, carried suitcases so heavy the handles broke. The gold was shuffled off in blanket rolls and packing crates, canvas sacks and glittering fruit jars. Five thousand people crowded the docks at Seattle to catch a glimpse of both the gold and the golden princes who looked more like paupers as they wavered under the weight of their treasure.

All that gold had been dug from an unlikely stream seventeen hundred miles up the Yukon River, where natives had fished for salmon by pounding stakes to hold nets. Their name for the place—*Throndiuck*—meant "hammer water." The anglicized version of that name would soon roll off the tongues of an entire nation—gilded with opportunity, promising one final fling into madness, a last grasping excuse for bank clerks and shoe salesmen, farm hands and bartenders to quit their jobs and rush off to that golden stream, the Klondike. Fifteen hundred people left Seattle within ten days and a hundred thousand more began making preparations.

If the language of gold was hyperbole—that larger-than-life exaggeration we all hope to believe—the Klondike was the hyperbole of gold rushes. The mountain passes would prove higher, the trails crueler, the climate more deadly, the boom briefer, and the misery more certain. There would be more suffering, more frustration, anger and despair, more absurdity, initiative and ingenuity. And the gold, ah the gold, there was more of it in a smaller space than anyone could imagine.

And no one was more surprised by this fact than those who were on the scene in 1896. Small strikes had been made along the Yukon River for a decade, at places like Fortymile and Circle City, and a handful of men had gambled on making their fortunes in the far north—men like Joe Ladue who couldn't marry his sweetheart back in New York because her parents didn't think he could support her in style. And Clarence Berry, who had a fruit farm in California before he was wiped out by the 1893 depression. He had hauled his bride on a sled over the high passes and down the frozen rivers, looking for a new start in 1895. There were others, there, who were misfits in the more civilized world, like George Washington Carmack whose sole ambition was to become an Indian.

One man, years before, had studied maps of the continent and deduced that gold cropped up all along the Divide and should intersect with the Yukon River somewhere in Alaska. Being remarkably pragmatic for a man of vision, he went to see for himself. The spring of 1883 found him at the mouth of the Yukon in the Bering Sea. He began edging his way up the mighty river in a tiny steamboat, and if Arizona had seemed ominously wild to Ed Schieffelin, it was nothing to this empty north land. By late summer he had covered a thousand miles and found a little bit of gold among the mossy stones. But if a star was shining down on the distant salmon stream as it had shone on the hills of Tombstone, the light was too bleak for Ed Schieffelin. With frost nipping the air, he turned his little boat around and left the north forever. Schieffelin died a year before the world found out that he should have kept looking.

Robert Henderson, who belonged to the same church of gold as Schieffelin, had more faith. Tall and angular, the Nova Scotian had dreamed about gold since childhood, and in 1894 was testing the gravel of the Yukon's tributaries. With a grubstake from Joe Ladue, Henderson spent two years searching between the Indian and *Thron-diuck* rivers. In the summer of 1896 he raised promising color on a creek he named Gold Bottom, and in the sharing tradition of sourdoughs, went off to spread the good news. One of the first people he told was George Carmack, who was fishing at the mouth of the *Thron-diuck* with his Tagish Indian wife and her two brothers.

With mild curiosity, Carmack and his two Indian buddies, Skookum Jim and Tagish Charlie, wandered up a tributary called Rabbit Creek, and while only half looking for gold found an outcropping of bedrock that yielded a shotgun shell full of nuggets from a few panfuls of gravel. Like the boys on Alder Gulch in Montana of 1864, George, Jim and Charlie danced a jig and clapped their hands and fell down in a heap while they dreamed of shiny new lives. In one of the small ironies so frequent in gold rush tales, Carmack, because of some perceived insult, didn't tell Henderson, across a tall ridge to the east, about his better find, and the gaunt partisan missed out while Carmack was given credit for the rush that followed to a creek renamed Bonanza on a river now called the Klondike.

Carmack created a veritable stampede by displaying his shell full of gold at Fortymile and emptying that small community overnight. The man tending bar in Bill McPhee's Saloon at Fortymile was Clarence Berry, the first to see Carmack's glistening shotgun shell. But by the time Berry secured a grubstake from his employer and dashed upriver to the Klondike, the best claim he could stake was Forty Above, numbered consecutively up and down Bonanza Creek from Carmack's discovery claim. Berry would later trade half of his Bonanza claim for half of a claim on the tributary Eldorado Creek, much to his good fortune.

Rumors were now drifting up and down the Yukon, and even over the mountains to the gold town of Juneau. The accumulation of empty boats at the mouth of the Klondike was reminiscent of the early San Francisco harbor, but the men rushing up and down the side creeks staking every spare piece of ground were simply going through the motions. Few believed there was really any gold there. Most left before winter, going back to where they came from. Some didn't bother to formally record their claims, much to their misfortune; others offered to sell out for two bits, and luckily had no takers.

The wifeless New Yorker Joe Ladue, aging in his northern gamble for wealth, rolled the dice one more time and left his trading post at Sixtymile to stake out a town site at the swampy mouth of the Klondike. He named the new town after a government geologist, and within two years Dawson would become the largest Canadian city west of Winnipeg. (The Alaska border, drawn along the 141st meridian, missed Dawson by fifty miles.) When Ladue stepped off the steamer *Excelsior* in San Francisco a year later, he was dragging a fortune in gold from the sale of town lots as well as a saloon and sawmill business. The press declared him the "Mayor of Dawson" and hounded him across the country to New York where he finally married his sweetheart, whose parents now lasciviously acquiesced. But Ladue's hard years in the north caught up with him, and by the next year he was dead of tuberculosis.

Back in the fall of 1896, no one yet realized how big the return would be. To reach a pay streak at bedrock, a miner had to tunnel his way down through fifteen feet of rock-solid permafrost by repeatedly building a fire at night, then mucking out a few feet of melted gravel in the short hours of daylight.

Clarence Berry, who could have been a linebacker for this century's *49ers*, was the first to hit bottom on Eldorado Creek. Crawling out from his shaft with a panful of paydirt in early November, the results made everyone else start digging. Three ounces of gold in that first pan would turn into eight thousand ounces by spring, but there was nothing to buy with it.

A tent town that winter of '96, Dawson was locked in the sub-arctic cold and existing on starvation rations. "No other community on earth had a greater percentage of potential millionaires," historian Pierre Berton noted, "yet all of its citizens were living under worse conditions of squalor than any share cropper." By June, when the regular trade boats arrived, eighty of those starving "potential millionaires" hopped on board with all the gold they could carry and escaped to Seattle and San Francisco where thousands would watch them and their three tons of gold disembark, and a hundred thousand would rush madly back toward the place they had departed.

Of course, there was no easy way to get to the land of hyperbole, but few in the stampede realized this. Several thousand attempted the all-water route, a simple boat ride to sudden wealth by steamer along the coasts of British Columbia and Alaska, around to the Bering Sea, and up the mighty Yukon for seventeen hundred miles. Nine-tenths were frozen in short of their goal, stranded in the wilderness through the seven-month winter.

There were others who believed the shortest distance between two points must be a straight line, and jumped ship at Prince William Sound or Yakutat three hundred miles due south of Dawson, only to be confronted by the sprawling and treacherous glaciers of the coastal mountains. They went snow-blind, or fell into yawning crevasses, were buried in blizzards, had no firewood to cook their food, and the few who made it past the summit of the mile-high sheets of ice lost everything in the rapids of the rivers on the far side, or lost heart when they looked at row after row of timbered ridges between them and their glittering goal.

Several trails led out of Edmonton, Alberta, the all-Canadian-boostered route. The shortest overland trail from Edmonton, along the Peace River, was seventeen hundred miles long through dense timber and boglands, across torrid rivers and mountain divides. Only a fifth of those who tried this route arrived in Dawson, spending two winters on the trail. The others who didn't turn back died of scurvy or gave up in frustration and shot themselves. The water route from

Edmonton, down the north-flowing Slave and Mackenzie rivers, wandered twenty-five hundred miles, crossed the Arctic Circle, and joined the Yukon well below Dawson. Half the people on this route succeeded, but arrived at the Klondike two years too late.

The most popular route to the Klondike involved a boat ride up the coast through Lynn Canal which parallels the thin strip of southern Alaska's panhandle. Near the end of the canal was the new town of Skagway, ruled by the scalawag Soapy Smith who had honed his art of knavery in the Colorado boom camps. Smith thought he was doing the *cheechakos* a favor by relieving them of their money here, where they could turn back, thus saving them from the struggle and disappointments sure to meet them further along.

From Skagway, a forty-five mile trail crossed White Pass and ended at Lake Bennett on the headwaters of the Yukon, a sensible trail that pack animals could follow across the daunting Coast Mountains. But the trail was sensible only in theory, as E. Hazard Wells, a correspondent for the *Cincinnati Post*, found out when he arrived at Skagway in early August of 1897. The trail had quickly deteriorated under the first onslaught of pack horses, and was soon impassable. The throng of would-be miners called a mass meeting to collect money and laborers for repair work.

As Wells described it, "a large body of men, perhaps 175 in number, under the command of the provisional executive committee, including your correspondent, started into the mountains to rebuild the old trail, but the gold hunger grew so keen that after a few corduroy bridges had been thrown across marshes and streams the attempt was abandoned and a general helter skelter ensued."

The trail over White Pass became "jammed with horses, goods and men, and catastrophes are occurring every hour," Wells observed. "This morning it is reported that over twenty pack animals have pitched off of bad places along the mountainsides. ... Every day a number are maimed or killed." The trail wound up and through the rushing Skagway River and over five mountain summits, with perilous switchbacks up one side, ten-foot boulders down the other, and mud holes in between that would swallow a horse, pack and all.

"The horses died like mosquitoes in the first frost, and from Skagway to Bennett they rotted in heaps," wrote eyewitness Jack London.

Packers were paid a dollar a pound to move freight across the trail, but even the best couldn't save their horses from the hazards of the trail. Legs snapped between rocks, horses drowned in the rivers, or starved from lack of feed. Sometimes men shot the horses that could go no further, and sometimes they left them to be trampled to death. Of the crowds he met Jack London wrote, "their hearts turned to stone—those which did not break—and they became beasts, the men on the Dead Horse Trail."

E. Hazard Wells also commented on the inhumanity that marked this route. "Two old men, gray-haired and bent, sat crying beside a muck hole. In it were their two horses, mired and helpless. Hundreds of men passed them a little to the right, and not one offered assistance. Not even a look of pity or sympathy was bestowed upon the unfortunates. The craze for gold had steeled the hearts of those who were once human beings."

Only a few hundred managed to get their outfits across White Pass and to secure a boat on Lake Bennett before the Yukon froze solid. Those that made it, including correspondent Wells, were traveling light and arrived at Dawson only to find the residents there evacuating in a panic because the supply boats trying to make a second trip upriver from the Bering Sea had been frozen in short of their goal. All the gold in the world couldn't buy what wasn't there, and the new arrivals, who had sacrificed much to be first in the rush, had to leave their goal as soon as they reached it.

"Of all the strange experiences of my life, this Klondike trip beats the record," Wells wrote. "After a cold, tempestuous journey down the Yukon, I reached Dawson City only to find a wildly exciting condition of affairs. Provisions were found to be running short in the town. ... Three days ago [October 1] the resulting panic reached its height among the 7,000 people here." Many fled downriver, hoping to find the supply boats or some other post with food. "Those who have departed ... by every known form of primitive conveyance may reach their destination before the river shall freeze up, but to make sure of doing it they will have to be expeditious." Wells himself fled in December, mushing up the Yukon by dog team and arriving in Skagway thirty-five days later.

Skagway had consisted of two wooden buildings and "half a hundred tents" when Wells left it the previous August. By the end of September, when rain and early snows made the White Pass trail even more impassable, the population had swelled to seven thousand residents. But another trail six miles up the canal was coming into its own with the first snows of winter.

Based out of the tent town of Dyea, the Chilkoot Trail climbed over thirty-five hundred feet, with the final ascent so steep no pack animals could cross it. Human power alone would lift forty thousand tons of supplies across the abrupt summit of Chilkoot Pass. The North West Mounted Police maintained an outpost at the summit to make sure that each man crossing into Canada had at least a thousand pounds of food and another thousand pounds of tools and supplies. Every man crossed the pass an average of forty times, portaging his gear fifty pounds at a time in stages, up and over and back down Chilkoot.

"It took the average man three months or more to shuttle his ton of goods across the pass," Pierre Berton noted, "and by that time the word 'stampede,' which connotes a thundering herd running untrammeled across an open plain, seemed a cruel misnomer." The trail began with a short wagon road up the Dyea River, narrowed through a two-mile canyon to Pleasant Camp, then started the climb to Sheep Camp at the edge of timberline, the last source of wood fuel. Fifteen hundred people were coming and going from Sheep Camp all winter. Beyond, the snow-blanked wall of Chilkoot towered, with only two resting places up the four-mile summit—Stone House and the Scales. The final, nearly vertical face had a single line of steps cut into the ice. "From first light to last, the line was never broken as the men who formed it inched slowly upward," Berton described, "climbing in that odd rythmic motion that came to be called 'the Chilkoot Lock-Step.'"

Seventy feet of snow fell on Chilkoot during the winter of 1897–98, burying two levels of stacked supplies cached at the summit. The Mounties had to keep sentries shoveling out the door of their shack so they wouldn't be buried alive. An avalanche, sweeping off the peaks in April killed fifty-six of the upward-inching stampeders. Will Shape stayed in camp that day because it had been snowing hard for nearly a week. "Some of the foolhardy ones did go out in the storm,"

Will wrote in his journal April 3, 1898, "and this morning something like fifty or sixty lost their lives in a big snowslide just above Stone House. … The snow on the mountains without warning came sliding down the slope on both sides of the gulch, packing tighter as it rushed along, until the bottom was reached and there it was piled up twenty to thirty feet deep. The unfortunate men caught in the trap had no show for their lives."

Will Shape had contracted gold fever in the epidemic that swept the nation during the summer of 1897. The thirty-year-old left a wife and two sons in New York City, hopped a train to Seattle on August 24 with a partner, George Hartmann. George was a cabinet maker from Brooklyn, and Will had trained as a goldsmith in Zurich. They bought their requisite sleds, picks, pans, and shovels, flour, beans, bacon and desiccated vegetables in Seattle, and shipped out for Haines Mission with four head of cattle on September 8. Their original plan was to pack their outfit on the cattle (two oxen, a steer, and a Jersey bull destined for the slaughter house) over the Dalton Trail just west of Chilkoot. But the Dalton Trail proved as hazardous to cattle as White Pass Trail was to horses, and one by one they lost their beasts in bog holes. Turned back finally by snow on the passes, Will and George retreated to the Chilkoot Trail, and began to pack on their own backs and on sleds their picks and pans, and beans and bacon out of Dyea on December 17.

It took them a month and a half to haul it all to Sheep Camp, fourteen miles from Dyea. Will met a tired looking man there one day "who was hauling a heavy load over the trail. He had his wife and little daughter with him. When I passed him, he turned to me with a look of despair overspreading his features and said, 'God Almighty keeps his gold in a mighty safe place, don't he?'"

The month of February was spent hauling loads up to Stone House, a huge boulder on a ledge beside the trail. "This day we hauled 600 pounds, in three trips, to Stone House," Will wrote February 12. "It takes two and a half hours for the round trip of three miles, on account of steep grades." By April they had succeeded at piling all of their cargo on the summit of Chilkoot where the Mounties checked the contents and charged them an import duty of $47.25 for their thirty-six hundred pounds of goods.

When Will and George finally reached the shore of Lake Bennett, it had taken more than four months to move forty miles. From Bennett it was a simple boat ride down to Dawson, but they didn't have a boat. And they didn't have lumber to build a boat. Like twenty thousand of their compatriots (for the White Pass and Chilkoot trails converged at Bennett), George and Will had to fell trees, drag them to the lake shore, and saw them into boards. "Neither of us had any experience in boat building, but we observed how others went at it then got to work ourselves," Will wrote.

The shoreline resounded with the crashing of trees, the ripping of saws and the cursing of sawyers. Logs were rolled up on a platform, lines snapped into board widths, and one man stood on top guiding the blade up while one underneath pulled the saw down along with a shower of sawdust on top of him. "It was no easy job for inexperienced hands to keep the saw straight on the line, and if once you run off, it is difficult to work the saw back again," Will observed. It was here that many of the partnerships began to fray. Men who had stayed together across the passes now grew uncontrollably angry at each other and demanded a division of outfits down to sawing their boat in half so each would get an equal share.

Pierre Berton tells the story of two former bank clerks who had been "friends from childhood, had gone to school together and worked side by side in the same bank as youths. They became so inseparable that, rather than be parted from each other, they married sisters. Yet the whipsawing turned them into enemies so insensate that when they decided to divide their outfits, they insisted on cutting everything in half. So bitter and obdurate was their enmity that, rather than divide twenty sacks of flour into two piles of ten sacks each, they persisted in sawing every sack in two. Then each set off with his twenty broken halves, the flour spilling away from the torn and useless containers."

Will and George built their boat uneventfully. Their undoing would come further along, as the two set off downstream on the first of June, 1898. "A hundred boats are within sight now," Will wrote as they sailed across Lake Bennett, "all bound for the Yukon River." Their own boat was twenty-two feet long with a flat bottom, six-inch keel, five-and-a-half-foot beam and a sail sewn together from canvas sacks. Down

Tagish Lake then Marsh Lake and into the Fiftymile River they floated in four days until they hit their first rapids at Miles Canyon and missed the pull out where boaters stopped to reconnoiter.

"The river narrows down suddenly and the current is very swift.... In three jumps I reached the sweep-oar in the stern of the boat. George remained at the oars and I shouted to him, 'We are in for it—let's try to keep cool and do the best we can!' (But the hardest part of it was to keep cool.) All this transpired in a few seconds. In spite of all the danger and excitement I could not help laughing when I saw George gazing at me with pale countenance and dumb as an oyster. Already we could hear the roaring of the waters in the canyon, some distance ahead. I was badly frightened myself, but for some unknown reason could not control my feelings and fairly shook with laughter."

Several sharp turns in the river obscured their view ahead, and at the canyon's mouth the roar of water was deafening. "My eyes were glued to the narrow opening in the rocks ahead and I steered straight for it, at the same time directing George how to row. Something is surely going to happen—I feel it coming. Into the narrow passage our boat shot, like a flash. On either side were jagged rocks—we must keep to the middle, right in the worst of it. How that water did tumble and foam and roar! There are two of these narrow shoots here, a wider bit of water connecting them. It is very difficult to keep the boat on a straight course. All of a sudden it made a lurch and swung around so fast, I could not prevent it. With a mighty tug at the oar I attempted to swing it into position, when, to our horror, the oar snapped in two."

Will grabbed a spare steering oar, but already the boat had spun around and was going down backward. Will got it turned right again in the wider channel and the two rowed wildly for the bank, barely missing a wrecked raft and crashing to a stop on rocks twenty feet from shore. They patched up their boat and hired a pilot for $25 to steer them through the Whitehorse Rapids that followed.

On June 6, Will commented that George was being disagreeable. "He continually worried over the outcome of the trip. I let him have his way in everything, just so as not to get into any arguments. He even threatened me, but I realized that the worry and excitement had a bad effect on him and asked him to be reasonable."

By June 10 they had reached the Thirtymile River, dodging rocks in the swift current, and three days later the Lewis River where the two again had a falling out. "[George] became impertinent, abusive and there was some talk of parting at the next landing place. He had his own way too much and could not bear the slightest interference in anything, without becoming ruffled. We camped at 6:30 and came to a thorough understanding, so as to avoid any complications in the future." Floating finally down the Yukon, on June 17 they reached the Stewart River which they decided to follow up and prospect, joining with two Englishmen who had a smaller, more navigable boat.

They towed the little boat up the Stewart, clawing their way along the bank, crawling across high bluffs, and wading often in the cold water. On the 9th of July, George and Will got into a fist fight. "I need not minutely describe what happened," Will wrote, "suffice it to say that he will never forget the severe drubbing I gave him. Even then he threatened to throw a big rock at me and only after repeated warnings that he would get a worse dose of the same medicine did he desist from his cowardly purpose. It was understood that when we reach the McQuesten River, our outfit would be divided equally and each could join whatever party he pleased."

The four prospectors continued upriver, but when they stopped again to cook a meal, Will turned around and "to my surprise I saw George standing about ten paces away, with the Winchester rifle leveled at me, threatening to shoot if I approached him. ... He acted like a crazy man and insisted on a fair division of the outfit, whereupon he was informed that the agreement entered into would be carried out to the letter, so far as I was concerned and that I would also take good care that he lived up to it. For just a moment I felt sorry that my gun was not handy, for he had me at a terrible disadvantage; but realizing that he was laboring under intense excitement, I simply insisted that he place the rifle back into the boat. This he bluntly refused to do, keeping his finger on the trigger and I really believe he would have shot me had I approached him." One of the Englishmen intervened and convinced George to put away the rifle, and the four continued on to the McQuesten River, which they reached July 12.

"The first thing on our minds was the dividing up of the outfit, as per agreement. This was no easy matter, as we soon discovered. Money and Cheatham [the Englishmen] were to divide it up—we to abide by their decision. With the provisions there was no trouble—but when it came to the tools and other implements, they knew not what to do. Surely they could not give one man a plane and the other an ax, or vice versa. So the judges decided they could not divide it equally and do justice to both. Finally Money took me to one side and asked me to reconsider the matter, while Cheatham argued with George. The result of it all was that we did not part."

For two more days they prospected haphazardly upriver, then turned around and floated down to a river bar on the Stewart that had shown the most promise. They built two rockers and began washing gravel. The two New Yorkers were gold miners for four days, the results of their labor $16 a piece. Without even bothering to float fifty miles down to Dawson, Will and George went home. Backtracking upriver, then cutting across on the Dalton Trail, they were back in Seattle on September 13 and soon left for New York. It had cost them $1,000 each, and fourteen months of their lives. "While it was not a profitable [trip] financially," Will Shape wrote in the end, "I shall never regret the experience." I'm not sure George Hartmann said the same.

While Will and George were rushing out of the Yukon River valley, Martha "Polly" Purdy was rushing in. She had caught the fever a year later than most, leaving a fine home in Chicago and two school-age sons. She was also leaving her husband, though that hadn't been her intention. Will Purdy, a paymaster for a Chicago railroad, had decided a jaunt to the Klondike might be rewarding, and with backing from wealthy relatives he set off for Seattle in June of 1898, accompanied by Polly, his wife of ten years. But at Seattle Will Purdy got cold feet and decided to go to Hawaii instead. Polly, in turn, decided that Will was undependable, she wouldn't go with him, and that she never wanted to hear from him or see him again. And she never did.

Polly convinced her brother George to accompany her to the Klondike, and off to Dyea they sailed to assault the Chilkoot Trail. While the winter trail was a blank sheet of snow and ice, in summer the pass was a tangled heap of house-sized boulders. Polly hired packers to haul their ton of gear to Lake Bennett for $900, and began the hike unencumbered. She spent the first night in a hotel at Sheep Camp, and started the next day over the summit. "As the day advanced the trail became steeper, the air warmer, and footholds without support impossible. I shed my sealskin jacket, I cursed my hot, high buckram collar, my tight, heavily-boned corsets, my long corduroy skirt, my full bloomers, which I had to hitch up with every step. We clung to stunted pines, spruce roots, jutting rocks. In some places the path was so narrow that, to move at all, we had to use our feet tandem fashion. Above, only the granite walls. Below, death leering at us."

Near the summit her foot slipped and she fell into a crevice, cutting her leg on a sharp rock. "I can bear it no longer," she says, exhausted, afraid and in pain; "I sit down and do what every woman does in time of stress. I weep. 'Can I help you?' 'Can I help you?' asks every man who passes me. George tries to comfort me but in vain. He becomes impatient. 'For God's sake, Polly, buck up and be a man! Have some style and move on!' Was I mad? Not even allowed the comfort of tears! I bucked up all right," and to the summit she climbed.

Camped at Lake Lindemann, waiting for her boat to be built by contractors at Bennett, she found bear tracks outside the cabin door one morning and afterward insisted on sleeping with a loaded gun beside her bunk. "One night I was aroused by a stealthy-creeping rustling noise," Polly recounted. "I got up quickly, put on my slippers, grabbed my gun, and ran across the dirt floor. … I crept to the window, raised my hand to remove the canvas screen, and let out a piercing shriek. Up my sleeve and out of the back of my gown was running the animal that I feared most of all in the world—a mouse. Did I voice that fear too—with shriek after shriek, arousing the whole camp! For days it was a standing joke. One old-timer looked me over speculatively, saying, as he chewed his wad of tobacco, 'Walked over the Pass. Goin' through the rapids. Campin' a long ways from home. Pretty rough life. Ain't afraid of nuthin' but a mouse. Lordy! Wimmin is queer.'"

Polly set off three weeks later in her $275 boat. At Tagish Check Station, one of several where the Mounties kept track of the stampede, Polly was informed that 18,000 men had passed that point since the previous May, and she was the 631st woman. Included in that total were two women newspaper correspondents (a new trend in both journalism and gold rushes), and the daughter of a Scranton coal miner, named Belinda Mulroney, who would become the richest woman in the Klondike. Not included were the dozens of dance hall girls who had steamed up the Yukon to further steam up the dozens of Dawson saloons.

It took Polly a mere twelve days to float down to Dawson where her brother and several partners built a log cabin in Lousetown, an unfashionable suburb across the Klondike from Dawson. The cabin had one large room with a small corner room partitioned off for Polly. But the prospects in Dawson that fall weren't promising. All of the gold-bearing creeks had been staked by mid summer, and crowds of stampeders wandered aimlessly up and down the city's muddy streets.

"It was as if the vitality which had carried these men across the passes and down the rivers, shouting, singing, bickering, and slaving, had been sapped after ten months of struggle," Pierre Berton explained. "For the best part of a year each had had his eyes fixed squarely upon a goal and had put everything into attaining that goal: but now that the goal was reached, all seemed to lose their bearings, and eddied about in an aimless fashion like a rushing stream that has suddenly been blocked."

By fall Polly reported that "everyone who could was hurrying out as fast as possible. A Mountie told me that there remained less than 15,000 of the 40,000 who had stampeded in, and before spring another 5,000 would be gone. ... As I watched the ever-departing stream I thought how only a few short months ago we, who had staked all and strained every muscle to get into the country, were now making every sacrifice to get out. I, too, must leave."

But Polly finally had to admit to herself what she had suspected for months. She was pregnant. As the daylight shortened ominously, she realized she couldn't leave. "I could never walk back over that Pass. Neither could I face the ravaging ordeal before me alone, helpless, most of my money gone. Life had trapped me," Polly lamented. She told her brother, and he broke down. "'I should never have consented to your coming,' he said. 'Father will never forgive me.' His helplessness to cope with the situation bucked me up," Polly admitted. "The winter days grew darker and darker, until it was continuous night. Endless days with no sight of sun. Deep blue nights with countless stars paling in the Milky Way. Cold, still, aurora nights, with red, gold, and green northern lights, crackling and swishing as they streamed from the dusky skyline to the very zenith of the heavens. Pale, green, moonlight nights, the Great Dipper, and, almost above us, the Pole Star, fixed, constant, comforting."

Unable to afford a $1,000 hospital bill, Polly delivered the baby herself. Little Lyman (named after his paternal great-grandfather) made quite a stir in the camp. "Miners, prospectors, strange uncouth men called to pay their respects. They brought gifts of olive oil (which by some miraculous mistake had been shipped in with foodstuffs), gold nuggets, gold dust. ... With tears in their eyes my visitors told me of their own babies so far away. The wanted to hold mine, to see his toes, to feel his tiny fingers curl in their rough hands."

Spring brought the trade boats again, and wildflowers along the banks which Polly loved to wander through and identify in the still evenings while the men minded the baby. The Fourth of July "was heralded in by a gunshot, one minute after midnight, and ten thousand Americans and Canadians paraded up and down singing alternately 'My Country 'tis of Thee,' and 'God Save the Queen.' The terrific uproar of unusual noises frightened hundreds of dogs that had never before heard such a racket, and they rushed madly up and down the streets or jumped into the river, to swim across to more silent places."

At the end of July Polly's father arrived unexpectedly to take her home. They steamed away upriver on a changed trail. A tram now carried travelers around the portage of Whitehorse Rapids, and a railway across White Pass transported them easily to Skagway, a calm town of ten thousand since Soapy Smith had been killed the previous summer in a gun battle, taking Skagway's most honest citizen with him in the end.

Polly and her father weren't the only ones leaving the Klondike in the summer of '99. Gold had been discovered in the beach sands of Nome, Alaska. "In a single week in August eight thousand people left Dawson forever," Pierre Berton noted. "And so just three years, almost to the day, after Robert Henderson encountered George Carmack here on the swampland at the Klondike's mouth, the great stampede ended as quickly as it had begun."

Berton also relates that "the statistics regarding the Klondike stampede are diminishing ones. One hundred thousand persons, it is estimated, actually set out on the trail; some thirty or forty thousand reached Dawson. Only about one half of this number bothered to look for gold, and of these only four thousand found any. Of the four thousand, a few hundred found gold in quantities large enough to call themselves rich. And out of these fortunate men only the merest handful managed to keep their wealth."

One of the handful was Clarence Berry, the Fresno fruit farmer who hauled his wife in by sled before anyone had heard of the Klondike. Berry took $1.5 million out of his Eldorado claims, struck it rich again in the rush to Fairbanks, bought oil property in California, and when he died in 1930 of appendicitis in San Francisco, he still had several million left.

Belinda Mulroney, the coal miner's daughter, had left a mail-order business in Juneau at the first hint of a rumor, crossing the Chilkoot in 1897. From the proceeds of her first cargo—silk, cotton cloth, and hot water bottles, she bought mining property, then built a popular hotel and saloon at the forks of Bonanza and Eldorado creeks. At the end of the rush she married a champagne salesman, and off they went to Paris, where, according to Berton, "they rode up and down the Champs-Elysées behind a handsome pair of snow-white horses, with gold-ornamented harness and an Egyptian footman, who unrolled a velvet carpet of brilliant crimson whenever they stepped out." Belinda died in 1967 in Seattle at age ninety-five, her fortune spent grandly.

Polly Purdy, who left the Klondike penniless, returned there in 1900. The North Star was her lodestar, she believed. She managed a sawmill her family established, and in 1904 married a Dawson lawyer. At age sixty-nine she became the second woman elected to Canada's House of Commons, representing Yukon Territory. Her son Lyman grew up where he was born, making pocket money by sluicing the dirt under floors of abandoned buildings.

But with gold it seems there are always more unhappy endings than happy ones. E. Hazard Wells described the luckless stampeders on the White Pass Trail in 1897. "At one place on the trail are camping a man and his wife. They are stranded in the mountains with their outfit, unable to proceed or retreat without abandoning it. The husband is a poor man and in despair. His wife spends her time in weeping. The couple have four small children in Seattle, Washington, left there with relatives. The entire family fortune has been spent upon the Klondike venture, and all is lost.

"Other men have mortgaged their homes; many have thrown up good situations; others again have sold out their businesses in the cities to embark on the Yukon mining speculations, and all are caught like rats in a trap on the Skagway trail. Ruin, desolation and despair brood over the region today, and many are the muttered curses, the tears and the vain regrets." Only gold could compel such voluntary misery.

But I wonder if it is fair to blame gold for what is really a crack in the human psyche, a fault between slip walls wanting to be filled with glassy quartz and the pale, glowing stringers of molten metal? The perfection of color, the immortality of form, the grace of simplicity, and the grandness of wealth.

Chapter 12

THIS IS WHAT the old, yellowed newspaper said, and so I went to look.

> Evidences of ancient mining have been discovered by U.P. Davidson and Louis Canard in the mountains on the head of the Little Popo Agie, which are creating a deal of excitement. The discovery consists of two furnaces built after the style of the old Mexican or Spanish furnace used for smelting very rich ores, and bears marks of having been built long years ago.
>
> One of the furnaces is of large size and was, in its day, capable of reducing a large quantity of ore, while the other one is small, and appearances indicate that it was used more as a testing furnace for ascertaining the value of certain new discoveries, much as the assay furnace is used at the present day, before submitting it to the purification process in the crucibles of the larger cauldron.
>
> Pieces of wood about four feet long are found laying about these furnaces so completely decayed that there is no mark of ax about them, yet it is plain from the uniformity of length and the position these pieces of wood occupy, that they have been cut. [As wood] in this climate will endure and remain in an almost perfect state of preservation for a period of twenty-five or thirty years, it is evident this cutting has been done much longer.
>
> In the near vicinity of these furnaces some very fine ore has been discovered, but the workings from which these furnaces were supplied have not yet been found.
>
> *Fremont Clipper*, December 19, 1888

The Little Popo Agie (pronounced po-po´sia, meaning "beginning of the waters") heads in a glacial cirque eighteen miles north of South Pass at a lake named Christina by one of F.V. Hayden's geologists in 1877 who claimed that it surpassed "in beauty of surroundings any we saw on the eastern slope of the range." The river runs some forty-five miles to join the main Popo Agie, and drops forty-five hundred feet in the process.

I must guess in my search that "on the head" would be somewhere in the first ten miles. Beyond that the river drops into an impassable snake-and-poison-ivy-infested canyon, and comes out five miles later on what, in 1888, was the main road from South Pass to "Lander's City."

Narrowing down the search is the first of my problems. The second is authenticating the original discovery. If these furnaces did indeed exist, who built them, and when? The who question seems less problematical than the when. Spaniards in the New World used primitive smelters in the rich silver mines of Mexico from the 1600s up to the beginning of the twentieth century, and their mining technology moved northward with them as they colonized Arizona, New Mexico and California. Otis Young, an authority on western mining, states that "dry-masonry edifices obviously used in processes involving combustion are frequently found about abandoned mine workings in the Southwest."

They look like squat, freestanding fireplaces with a small fire box where charcoal and forced air could heat crushed ore enough to smelt out metals. Few Americans were familiar with Spanish mining techniques prior to the California rush; they had no need to be. But U.P. Davidson, who found the furnaces on the Little Popo Agie, would have recognized their Spanish origins. He was an "old Californian" who survived an 1849 crossing of Death Valley while others on the route died that year and were consumed by their fellow travelers.

If I grant that these furnaces were built by miners of Spanish descent, I must still guess when. I have seen wooden relics here a hundred years old and still recognizable. That would push the date back to 1788. If we take the reporter's estimate of "perfect preservation" at twenty-five years and double that for decay, it would place us in 1838, well within the fur trade era. That is a fifty year window to look through.

Beyond that window I see Juan María Antonio Rivera, who traveled north from Santa Fe in 1765. It was his diary that Escalante followed in 1776, long thought to be lost but recently discovered in the archives of Madrid. Rivera found what he thought was gold ore in southwestern Colorado, near what would become the mining town of Durango. When Rivera reached the Colorado River near what is now Moab, the "Yuta" Indians told him a curious thing. Beyond the river six days journey were "some bearded white men dressed in armor with metal hats."

Six days north would put them in the Uinta Mountains, site of the "Lost Rhoades Mine" where Brigham Young was said to get the gold for his newly minted Mormon coins.

Escalante thought Rivera's bearded Spaniards might be the *Yutas Barbones* who lived along the Sevier River of western Utah. They grew heavy beards, and wore bird bones through their noses, and when Escalante met one "venerable" old Ute there who lived alone in a hut, "his beard was so thick and long that he looked like one of the hermits of Europe."

But as late as 1811, Don José Rafael Sarraceno, postmaster of New Mexico, journeyed north "in an effort to locate a Spanish settlement which the Yutas have always asserted lay beyond their territory, supposedly completely surrounded by wild Indians. After having traveled for three months, he was finally stopped by a large river. Among the Indians living there he found many articles manufactured by Spaniards such as knives, razors, and awls; he obtained the same information there, that the makers of those articles lived across the river."

I can find nothing in the official records about mining on the northern frontier, which makes it nearly impossible to document the rumors of old Spanish mines in Colorado and Utah. They might be silent for several reasons. Unlicensed trading with the Utes was banned by successive governors at Santa Fe who thought such trade incited hostilities. But the arrest records of dozens who went north despite the ban suggest that the laws were ignored frequently and repeatedly, and that illegal traders needed to keep their mouths shut about what they'd been up to in the three to six months they'd been gone.

Another reason illicit mining demanded secrecy was that the King of Spain remanded a royal fifth of all profits. If you didn't advertise your gold, you wouldn't have to share it. It was reason enough to keep mining activities out of the royal record, but I still have to wonder if anyone can keep a secret when gold is concerned.

The unofficial record, represented by folk history, locates old Spanish mines all over Utah and across southern Colorado. Some of them appear (as vague x's on the map) in highly unlikely places, but others are in areas that became profitable mining districts. And there are some that are supported by contemporary records which lend credence to the rumors. A reporter for the *Wasatch Wave* of Heber, Utah, recorded his visit in 1897 to an old mine tunnel in the Wasatch mountains that was in the process of being dug out by local ranchers. "The work of cleaning out the tunnel is done mostly with shovels, laying bare the top and sides of solid stone, which plainly demonstrates that the implements of man have been used in the first excavation." The tunnel (adit is the correct mining term since tunnel implies an entrance and exit hole) had been driven "twenty-five feet into the solid rock following a vein of ore from the surface, which is reported to return very good assays in gold."

Down the ridge half a mile was a five-foot slab of granite "covered with peculiar looking hieroglyphics cut into it. . . . One of the figures is of a man with hands thrown up as though suddenly surprised; another is what we would call that of a burro or pack mule; another a half moon, and there are a number of others, while perfectly visible, we were not able to decipher their meaning." The discovery was twenty-five miles south of the Park City mining district, and as the reporter commented, "all early settlers in the valley are familiar with the story that at a time some eight or ten years before the first settlers here, there existed valuable mines in the surrounding mountains which had been worked by Mexicans."

William Palmer asserts that "Little Salt Lake, in southern Utah, was known to white men nearly a half century before James Bridger discovered Great Salt Lake. Spanish priests, traders and soldiers of fortune traveled the Old Spanish Trail every year, and there are evidences that Iron County was well prospected for minerals. After the Mormons settled Iron County in 1851, they found many little prospect holes in the hills." They also found an old adit in Coal Creek Canyon that an intrepid youth crawled into despite bear tracks around the entrance. In the mine he found "an old double-pointed Spanish pick" which was given to the L.D.S. Museum for display. In 1896, the nearby Escalante mine sank shafts on twenty-foot-wide native silver veins.

Spanish miners would have had good pickings in Utah, where more than 130 organized mining districts have been discovered. But Utah never had a gold rush on the scale of other western states for several reasons. Placer gold, which can make or break a rush, was notably lacking in most of the mineral districts partly due to the dry climate, but also because the Tertiary deposits were primarily copper or silver ores with gold as a by-product. Bingham Canyon, discovered in 1863 at the southeast end of Great Salt Lake, was the state's biggest placer gold producer, but open-pit copper mining proved the real bonanza at Bingham. On the opposite side of the Oquirrh Mountains, the Ophir district turned out $92 million in lead-silver ore, and the Mercur district produced $25 million in silver and gold.

Southeast of Utah Lake, Eureka would anchor a series of booms in the 1870s throughout the Tintic district where rich surficial silver veins returned $400 million. It is possible that the silver was discovered much earlier. The *Salt Lake Herald* reported on May 19, 1871, that "yesterday a miner came to town with some very rich ore taken from an old mine, which when discovered still showed chisel marks and other evidence of having been worked in by-gone days. He stumbled onto the old shaft by accident, within ten miles of Tintic, although he is very reluctant to say just where. The ore he brought in assays at more than $6,000 to the ton."

I don't know how to reconcile the differences here between rumor and lack of facts, the discrepancy between folk history and official history whose story is always incomplete. What do I make, for example, of the curious fact that Astorian Robert Stuart, when he crossed the upper Green River in 1812, called it the Spanish River? Did he know there were Spaniards living downstream? South Pass was an easy five-day journey from Escalante's 1776 crossing of the Green near Vernal, Utah. Did the illegal Spanish traders venture much further than we have assumed? It is also possible that Spaniards came from the east instead of the south, for Spain controlled the Louisiana territory west of the Mississippi for forty years between 1763 and 1802 before Napoleon took it back and sold it to the United States.

Wherever the Spaniards went, they must have been looking for gold, because the Spanish link to gold in the New World was long, looping, and hot-forged. It began with Spain's sponsorship of Columbus, and was followed by a decree from the Pope in Rome granting Spain all newly discovered lands west of about 40° west longitude, and Portugal all new lands east of that line (which is why Brazil, whose coast spills over the line, has Portuguese as its official language while the rest of Central and

South America speak Spanish). Given free rein to conquer and "convert" New World natives to Christianity, Spain quickly expanded its foothold in the Caribbean and began exploring the surrounding mainlands "for God, for glory, and for gold," but not always in that order.

The gold of the West Indies was limited in amount and crude in workmanship, a disappointing bounty for the conquering armies. But rumors of golden cities on the continent drew many a conquistador inward and onward like a desert mirage. They were expeditions remarkable for their danger and hardships and lack of reward. In 1513, Balboa hacked his way through the jungles of Panama to see the Pacific Ocean, and Ponce de Leon discovered and named Florida. Cabeza de Vaca, shipwrecked on the Gulf Coast, wandered for six years across Texas, New Mexico and Arizona before finding a Spanish settlement on the west coast of Mexico in 1536. Hernando De Soto sought mythical gold in the southeastern U.S. only to die on the Mississippi, treasureless, in 1542.

And then there was Cortés, who believed in rumors of a golden city in the middle of Mexico. Hernando Cortés was born in 1485 in a country that had spent six centuries reclaiming their lands from invading Moorish infidels. Spain was a nation burning with religious fervor, and restless with a surplus of warriors who no longer had a war after the last Moorish stronghold of Granada was overthrown in 1492. The New World offered an outlet for this crusading energy, and an opportunity for wealth that Hernán Cortés found irresistible. He left Spain a decade after Columbus had shown the way.

Cortés was thirty-three years old when he began his conquest of Mexico in 1519. He had helped conquer the island of Cuba in 1511, gaining favor with the new governor of that island where he was given land and Indian slaves. Cortés bided his time, gathering wealth and allies. He had been a law student back in Spain, and he continued to be a student of the tricks and turns of human nature. When an opportunity arose to lead an exploring expedition to Mexico sponsored by the governor of Cuba, Cortés presented himself as the best man for the job. It was a choice the governor of Cuba would come to regret.

Cortés landed at Vera Cruz with 508 soldiers and sixteen horses. He built a small fortress and left a fifth of his men to defend the port, then sank his ships to prevent desertions. When the Aztec ruler

Montezuma sent gold to the Spaniards as a gift to speed them on their way back home, he miscalculated Cortés's true goal. The expedition changed from exploration to conquest, and Cortés started the first gold rush in the Americas.

Marching westward toward the Aztec capital 250 miles inland, Cortés and his small army would climb upward through thick jungles, and down through fertile green valleys, cross deserts of cactus and crest mountain ranges peaked with towering volcanoes, to finally view Tenochtitlan—what Cortés described as one of the most beautiful cities on earth. The Aztec capital was built in the middle of a great lake, with whitewashed stone glistening brilliantly against the blue water. A fortress of 300,000 people with causeways from the shore and drawbridges across the many canals, Tenochtitlan was a spectacle of great temples rising above the terraced houses, gardens, aviaries, and a sprawling market square.

Across the causeway and into the city Cortés marched on with his soldiers and horses. Thousands crowded close to watch the parade, and to the nervous soldiers, this unopposed entrance into the capital seemed all too easy. It was not as if Montezuma (the correct, though less familiar spelling is Moctezuma) didn't know Cortés was coming. The lands that the Spaniards crossed were not empty, but thickly populated wherever agriculture was possible, and the people were subjects of Montezuma either by choice or by conquest. From the coast, Cortés had fought and won stunning victories against the natives. They had superior numbers, but Cortés had superior weapons in his crossbowmen, musketeers, armored cavalry, and the mystique of things new and strange. Each army he conquered joined Cortés as allies against Montezuma, complaining that the Aztec king taxed them too heavily in goods and human life.

The gods of Aztec mythology had to be fed a daily diet of human blood. Their victims, young men and boys who were prisoners of war or tributes to the tax collectors, were laid by the priests on sacrificial stones, their chests sliced open with obsidian knives and their beating hearts ripped out and offered to the thirsty gods. The Spanish soldiers, who had been housed like royalty in a palace near the temples, witnessed this daily ritual and shuddered to think they might be next.

But something unearthly was working in their favor. A story in the Aztec cosmology told of a white god who would come from the east to take his rightful place as ruler. Montezuma wasn't sure if Cortés was this god, or an emissary of the god. He didn't want to anger Cortés, and he had tried all along the march to keep the Spaniards out of the capital by bribery and subterfuge. But the gold Montezuma sent only encouraged Cortés, and the armies he secretly ordered to oppose Cortés were defeated and turned against him. When Montezuma met with Cortés at the palace in an attempt at diplomacy, Cortés held him hostage in a quiet endeavor to capture the city with his own brand of diplomacy. Montezuma did whatever Cortés told him to do, and the strategy might have worked if Spanish ships hadn't arrived suddenly at Vera Cruz with instructions from the governor of Cuba to depose Cortés as self-proclaimed ruler of this New Spain.

Cortés marched back out the causeway, leaving 250 men to hold what he called Mexico City, back over the mountains and deserts, through jungle and valley to stage a night battle that captured the newly arrived ships and gained him 800 reluctant reinforcements. (As a side note, the commander of the ships, a Spaniard named Narvaez, lost an eye in the battle and was held prisoner at Vera Cruz for years. When Cortés finally turned Narvaez loose, he returned to Spain and spread such villainous tales about Cortés that he would lose his royal title to New Spain. But Narvaez was consistently a poor decision maker, and it was his expedition to the Gulf coast that was shipwrecked in 1528. Narvaez went one direction and was never heard from again, while Cabeza de Vaca went another into exploration history.)

While Cortés was away dealing with Narvaez, his captain unwisely ordered the massacre in Mexico City of a thousand unarmed Aztecs during a ceremonial dance, finally inciting the city into revolt. Cortés learned his soldiers were under siege and marched once more inland with his reinforcements. He crossed the causeway unopposed and tried to use Montezuma to quell the revolt, but the king was stoned to death by his own people. In a desperate retreat, Cortés fought his way back out the causeway, losing six hundred men in one night of fighting. The Spaniards had accumulated 700,000 pesos in gold (something like eight tons) during their eight-month stay in the city, melting down the royal treasure of beautifully rendered golden birds and flowers, jewelry and tableware into ingots they could divide evenly between themselves. They were soldiers of fortune in the most literal sense, and loyal only to a golden end, so when they retreated they tried to take their gold with them. Many of the six hundred who died drowned in the bridgeless canals, weighed down by the golden treasure they refused to leave behind.

Outside the city, Cortés mounted a final siege using 100,000 Indian allies. They spent three months fighting their way back in, destroying the city one house at a time, and slaughtering the inhabitants until the bodies were so thick they filled the streets. Within two-and-a-half years after landing at Vera Cruz in 1519, Cortés had replaced Montezuma as ruler of Mexico. He would rebuild the city and divide into feudal estates an area of land bigger than Spain. The treasure he captured would fire the minds of many a would-be conquistador, driving them inward and onward like a desert mirage. And one of the most ardent rainbow chasers was named Francisco Pizarro.

Pizarro was born in the same Spanish province as Cortés, and though related to Cortés through marriage, Pizarro was illiterate and lacked Cortés's diplomatic skills. He was a true soldier of fortune, beginning life as a swineherd, then shipping to the New World and proving himself a ruthless and hardened fighter. He was nearing sixty, with a small piece of swampland in Panama, when he heard rumors of a rich kingdom to the south and decided he had little to lose. What Pizarro lacked in finesse, he made up for with a desperate determination. He practically starved to death on a deserted island when he refused to return to Panama while his first armada sailed back for supplies and reinforcements. On his second journey down the coast of South America he crossed the Equator into waters no European had sailed before. On his third trip in 1531, Pizarro landed on the northern coast of Peru and marched inland to confront the mighty Incas with less than 170 men and sixty-two horses. The Inca ruler was relaxing at a hot spring high in the Andes Mountains, having recently won a civil war against his half brother, and he let the Spaniards come unopposed to visit him, even clearing out a town for their accommodation while he and his thirty-thousand-man army camped on the plain nearby.

Pizarro boldly invited the Lord Inca to sup with him in the town, and the invitation must have impugned the Inca's courage, for the great lord accepted the challenge and was carried into the town on a litter with only six thousand attendants, all unarmed. Pizarro had closely studied Cortés's conquest of Mexico, and when the Inca retinue paraded into the town they found it eerily silent and empty, with the Spaniards concealed behind walls with their cannons and cavalry and cold steel swords. Pizarro waved a handkerchief and the carnage began with such ferocity that the Indian guards were crowded against the adobe walls until one collapsed and they ran streaming out. The Inca lord was pitched back and forth on his litter like a capsizing ship on a ravaged ocean, and only Pizarro's intervention saved his life. Like Montezuma, the Inca Atahualpa became a Spanish hostage.

It was the Inca's own idea to ransom himself for a room full of gold. For months the gold was carried in, stripped from the walls of distant temples and piled in a glowing heap before the Spaniards' avaricious eyes. When the gold stopped coming, the pile was worth, by today's value, something like seventy million dollars. It wasn't expedient to turn Atahualpa loose as the bargain required, so the Spaniards held a mock trial, convicted him on various charges, and strangled him to death with a garrot. Pizarro picked a new lord Inca who would obey him, then captured the capital at Cuzco with little resistance. Pizarro ruled Peru for eight stormy years, looting all the gold he could find.

The material success of Cortés and Pizarro made the existence of other golden kingdoms seem not only possible but probable. On the sea coast of Columbia, conquistadors heard of a king far up the Magdalena River who was anointed in oil then sprinkled with gold dust. Pushed out on a raft into the middle of a sacred lake, he would take a Midas dive into the water to wash himself clean, while his subjects threw gold ornaments and precious stones into the lake as offerings. This story was so persistent that the Spaniards gave a name to the king—*El Dorado*—"The Gilded Man." But when the conquistadors reached the highlands of Bogotá, the gilded man was not there. Perhaps he was farther east across the Andes, they thought, and so they pushed on to look for him. They are still looking, and so are we.

It was another legend on the same vein that drew conquerors north from Mexico into what is now the American Southwest. In the Middle Ages seven Portuguese bishops fleeing from invading Moors crossed the ocean to found seven cities in a new land, according to the legend. The viceroy of Mexico (Cortés's successor), Antonio de Mendoza, heard rumors of a place with seven cities to the north of New Spain. It was a land rich in silver and gold, according to Fray Marcos de Niza who was sent north to verify the rumors, with towering cities of stone. If Fray Marcos actually laid eyes on the Seven Cities of Cíbola, he possessed a grand imagination. It is more likely that Marcos relied on native informants, who had a remarkable talent for telling the Spaniards exactly what they wanted to hear about the land of Eldorado, *más allá*, "a little further," just beyond the horizon.

Viceroy Mendoza chose thirty-year-old Francisco Vásquez de Coronado to lead the expedition to Cíbola, and both invested a large part of their individual fortunes in the gamble. But with the glittering example of Cortés and Pizarro still fresh in that year of 1640, and a golden mirage hovering on the northern horizon, how could they afford not to go? It was a seven-thousand-mile wild goose chase. Cíbola proved to be a group of Zuñi pueblo towns gilded only with sunlight, and the other pueblos up and down the Rio Grande basked in the same unredeemable light.

But *más allá*, across the prairie to the east, was a glorious land of gold called Quivira, according to a Wichita Indian slave held at one of the pueblos. He would show them the way, for he badly wanted to go home. So Coronado set out across the buffalo plains with his clanking army of five hundred soldiers plus natives, his thousand horses and mules, his herds of sheep and cattle which were a walking commissary. They were led astray by the Wichita slave, across the Staked Plain of Texas, so flat they could see the curve of the earth, so devoid of landmarks it was like a barren ocean of grass. When Coronado realized the Wichita's deceit (he wanted them all to become lost and die there), the native was placed in chains and the Spaniards turned north to cross Oklahoma and Kansas.

When they found Quivira along the Arkansas River of central Kansas, it was a collection of grass huts with no gold in sight. Eldorado, the Spaniards were told, was *más allá*, beyond the horizon. But Coronado had gone as far as he could go, and in August of 1641 he turned back. "So far as I can judge, it does not appear to me there is any hope of finding either gold or silver," Coronado wrote to Viceroy Mendoza. It was a dejected troop of soldiers of fortune who marched back to Mexico. They weren't the first to lose the golden lottery, and they wouldn't be the last.

There might be a risk in even buying a ticket to this lottery, for it would seem that a curse comes along with the gamble. We could attribute their bad luck to the times and circumstances, but Cortés, Pizarro and Coronado all came to a bad end due directly to their quest for gold. Cortés was stripped of his governorship of Mexico and banished from the capital. His later *entradas* to discover and claim new lands were disasters that wasted much of his fortune. Cortés died in Spain, much like John Sutter died in Washington, trying to recoup from the government the losses he had sustained.

Pizarro's death was as brutal and devious as his conquest of Peru, for he was stabbed to death in his own home by jealous rivals in 1541. Coronado faced a gentler death, but no less tragic. Back in pueblo land he had fallen from his running horse and was struck in the head by a flying hoof. He was carried back to Mexico on a litter, and never fully recovered. His power, his wealth and his mind slipped away quietly until he died in Mexico City at age forty-four.

I myself must contemplate this curse as I go in search of a lost Spanish mine. Even the Mexican miners in Utah's Wasatch, who left a self-portrait in stone—hands thrown up and pack burro beside them—were said to have perished, murdered by Utes and thrown down the mine shaft. So I begin the search in caution, down a steep slope to the canyon bottom where the Little Popo Agie drops inextricably through the staircase of rock that leads down to the Wind River basin. What I find is a cave, not a mine but an old Indian rock shelter cut out of the cliff by the boiling river. Its ceiling is smoke blackened, and flakes chipped off of stone tools litter the level, sandy floor.

On the limestone ridges above the canyon I find more of these flakes, plus the cores they were chipped from, beautiful agatized stone in purples and pearl gray, jaspers and chert. Signs of elk and mountain sheep are everywhere in groves of aspen, pine and fir, and dried leaves of balsam root crackle underfoot. It is not a likely place to find gold, though, so I continue my search upstream.

Bog and timber, bog and timber, the stair steps keep climbing upward in rugged leaps followed by flats full of lily ponds, sedges, and deep fishing holes. There is some placer gold in the Little Popo Agie, but it appears to have been washed in from the South Pass veins, and higher up the gold disappears. It seems an impossible task to do a thorough search of the lower half of this ten mile section, so the first week of September I decide to go straight to the source.

There are two ways in to the headwaters: by an axle-breaking wagon road turned Jeep trail, or by hiking trail. I pick the latter, shouldering my pack to begin the six-mile trudge uphill in a misting rain. It is more of the same, timber and bog, timber and meadow. At the end is Christina Lake, named for a geologist's sweetheart, and one of those spectacular views that makes the work of getting there worthwhile.

I pitch my tent in the lee of some scrub pines while the wind raises whitecaps on the ten-thousand-foot-high lake. It rains off and on through the night, a mountain rain that will not reach the sagebrush plains below. I look out in the morning at sparkling sunlight and wonder if I haven't wasted the part of my life spent indoors, especially since I've been rereading Thoreau's *Walden.* "Rest thee by many brooks and hearth-sides without misgiving. … Rise free from care before the dawn, and seek adventures. Let the noon find thee by other lakes, and the night overtake thee everywhere at home. There are no larger fields than these, no worthier games than may here be played." Gold is just my excuse, and on that reminder I set out looking for a gold mine.

The lake is more than three miles in circumference, a glacial cirque once filled with ice. The surrounding ridges reach ten, eleven, and twelve thousand feet above sea level, cobbled with loose rock and streaked by old snow in couloirs and cornices. A rim of meadow bands the shoreline below, and timber forms its own shoreline against the mountains where it can grow no higher. This is the beginning of an alpine world that stretches a hundred miles northwest along the crest of the Wind River Mountains, or it is the end of that world depending on which direction you look. Winter winds must siphon down the steep walls and roar across the frozen lake, for the eastern end is truly a "blasted heath." The white pines that have tried to grow there are gray, twisted skeletons, all boney limbs stretched leeward and frozen in a writhing scream.

Warmth and sunlight on this day make them otherworldly, and I am happy to leave them behind as I start around the lake by the south shore. There are old blaze marks cut into the trees, but the trail once marked here has fallen into disrepair. I trace its course around to the head of the lake, and there I find a name and date carved into an old pine tree, but grown over into illegibility. A horseshoe once propped in a fork has also been swallowed by rings of growth until only half sticks out in an unlikely appendage. Interesting but not evidential I decide, since trees here don't live much past a hundred years, and I am looking for clues twice that age.

I study with binoculars the walls of stone rising a thousand feet above and all around me, and I have to conclude that the possibility of a gold vein in this granite batholith is highly unlikely. The newspaper report was no more specific than "in the mountains on the head of the Little Popo Agie." Downstream from the lake, the river crashes through boulders of glacial moraine no more promising than the sheer walls that rock was plucked from. If there really were Spaniards smelting gold around here, I can only conclude they had brought the ore in and were hiding from the Indians. Off the well-beaten path at South Pass, the thin smoke from their furnaces would lose itself against the mountain ramparts, they would have wood to make charcoal and good hunting to keep themselves fed. How they found their way into this high bowl, I cannot imagine, but it seems unlikely they would be followed.

Pondering this theory, I continue around the lake, but get distracted by low-growing grouse whortleberry bushes. The tiny maroon berries have a bittersweet clove taste, and rummaging around in the forest undergrowth I also discover kinnikinnick berries, bright red and bell shaped, which taste like juiceless wild strawberries. Then I find wild blueberry bushes growing by the lakeside and I paw the berries off like a bear, grunting with delight as the afternoon slips by unnoticed. It is nearing twilight when I finally cross the outlet of the lake on my way back to camp. The water is still, and the quiet seemingly unbreakable. I stroll across the cobbled beach head, admiring the rounded stones, but I see something disturbing there. Wedged in the sand between the stones is the angular tip of an Indian spearhead, the flaked gray quartzite smoothed by waves. It makes me realize they couldn't have gone anywhere unnoticed, the Spaniards, with their armor and their cannons and their lust for gold. The deserts they crossed and the mountains they climbed were never as empty as they seemed. If the old Spanish furnaces are ever found again, I hope their makers left something behind to solve the mystery, some piece of armor, a hand-forged tool, or maybe a self-portrait in stone with hands thrown up in surprise.

There are other, more obvious signs of gold chasers here at Christina Lake. A dam built of stones and slabbed with boards seals the end of the lake, and a diversion ditch snakes off deviously into the timber to lift water from the Little Popo Agie drainage across the mountains to Rock Creek. Another ditch from a reservoir on Rock Creek hugs the canyon side above Atlantic City, falling away in broken flumes where the cliffs were too steep to blast into ditches. These diversions took twenty men two years to build, cost almost $70,000 and are now just dry, inverted monuments to the man who made them happen.

Emile Granier was a Frenchman, fifty-five years old when he arrived at nearly-deserted Atlantic City in 1884 and decided there was gold by the wagon load in the gravels under Rock Creek. Placer mining had always been difficult on the creek because the grade was shallow and the gold had settled under five to twenty feet of loose rock. Granier proposed to excavate the gold by hydraulic mining, which uses the force of gravity-pressurized water to move gravel in large quantities.

Hydraulic gold mining had become all the rage in western districts where simpler techniques like shoveling into sluices no longer paid off. It was first used in California in the 1850s, a purely American invention that washed down whole hillsides into oversized long toms where the gold was caught and the gravel swept away. The earliest attempts used canvas or rawhide hoses with small wooden nozzles, but later technology produced water cannons called Little Giants that could spew a solid stream hundreds of feet.

In places like Nevada City, California, where John Banks had witnessed miners digging "coyote holes" into gravel seventy feet deep in 1851, water cannons were soon excavating the same gravel with a simple twist of the jet. Some of the gravel had become cemented together, and had to be blasted back apart. One blast in Nevada County was exuberantly achieved with over five hundred kegs of powder, lifting a nine-million-cubic-foot mountain several feet in the air before it came down in pieces.

On the Yuba River, where ex-soldier Edward Buffum washed out his first flakes of gold in 1848, the mass blasting was community theater. "The windows of the houses in Sucker Flat proper are whitened with human faces," wrote a member of the audience in 1870. "They gather in groups on the side of the hill, below the store, and its large, back porch is jammed with men and boys. Wherever there is standing-room, giving a view of the mine, and thought to be reasonably safe, human forms may be seen, with eager, anxious faces." What chunks didn't land on the rooftops were carted off to stamp mills to grind the cement to dust.

But the discharge debris of the erosive hydraulic mining often caused problems downstream. Spring floods along the Sacramento washed the debris over farmers' fields in the Sacramento Valley ruining valuable crop land, clogging the way of river boats, and flooding yet again Sacramento and surrounding river towns. Farmers sued the miners and eventually won. In 1884 the California State Supreme Court ruled that "hydraulic mining constituted a general nuisance and that no person or corporation had the right to cover another's land with mining debris." This effectively stopped hydraulicking along the Sacramento, but it was still popular throughout other regions of the west that could collect the four essentials of success: sufficient water, correct technology, substantial capital, and the right terrain.

Emile Granier thought he had solved the water problem by building a reservoir above Atlantic City in 1885, but the reservoir ran dry by the end of summer. He then dammed Christina Lake and stole the water of the Little Popo Agie. The farmers far down that river got angry, and Granier had difficulty getting the legal right to his usurpation. By the time Granier had finished his dams and ditches, bought a sawmill to make boards for his flumes (one wooden ditch stretched five hundred feet on top of seventy-five-foot trestles), gained patents on four miles of Rock Creek, secured permits to use the water, and purchased mining equipment, he had spent more than $200,000 and six years without seeing a glimmering ounce of gold.

But Granier didn't seem worried about the money. He was backed by a syndicate of French capitalists, and when he returned to Paris every winter he must have assured them their golden reward was in sight. To the local *Fremont Clipper* newspaper he proclaimed, "of course I am anxious to see the work completed, but as for the amount of capital I have expended in this enterprise, I have no fear, and am perfectly satisfied that I will accomplish what I started in to do, surprise the world with the output of gold from my mines in the vicinity of Atlantic City."

The stocky Granier must have cut quite a figure in downtown Atlantic City, with his snappy foreign clothes, his Paris accent, and the money to employ sixty men at the only game in town. But he didn't seem to be a popular man. Granier wrote to the territorial governor in 1889 asking for troops to be sent because his life had been threatened by an unnamed resident of Atlantic City. "Permit me to beg your prompt attention to this matter as I am positively helpless here," he cried. It seems that Granier had secured title to the land under Atlantic City, but not the buildings themselves. He offered conditional deeds to the residents, but one unhappy denizen had cornered Granier on the street and demanded forcefully that the Frenchman either buy him out completely or give him an unconditional deed. When Granier realized the troops weren't coming, he paid off his extortionist and the crisis passed.

But the following year Granier's attention was again diverted by the residents of Atlantic City, who were afraid his dam on Rock Creek would fail and wash them all away. He spent the summer of 1890 shoring up his dam. Having solved the water and capital problems, Granier turned to the issue of technology. Rock Creek would not yield to traditional hydraulicking because it had no mountains of gravel to wash down, only a terraced stream bed between twisted black cliffs. Granier turned instead to a Hendy Hydraulic Gravel Elevator which used water pressure to carry gravel upward to a sluice box. Granier himself was uncertain whether the technology would work, and the question of its success is shrouded in history. Some say he made $200,000 in the three summers his operation was working, and just broke even. Some say he made nothing.

Emile Granier left for Paris as usual in the fall of 1892, but he never came back. It was the terrain of Rock Creek that finally caused his failure.

The grade was too flat, and the tailings piled up, causing back flooding. Rumors that Granier was thrown into debtors prison back in France are probably just the wishful thinking of an abandoned community. A *Wind River Mountaineer* article in 1907 reminisced about Granier, "who resides in Paris, France and is a very old man. It is questionable whether he will ever again visit the scenes where twenty years ago he built gigantic castles and painted every cloud of fleeting thought with the yellow tinge of gold." The one photo of Granier still extant shows the stocky man posed rakishly beside his ditch. His round face is circled by mutton-chop whiskers, his expression studiously grim. He is leaning on a cane and wearing a long canvas coat, polished leather boots and a funny little pith helmet. Put savanna in the background, change his cane to an elephant gun, and you can imagine him on safari.

You can still see the dams and ditches and broken flumes Granier left behind (good evidence of how long any diggings remain here), but all signs of his actual mining have been swallowed up by a more recent enterprise that actually proved Granier's theory of gold by the wagon load in the gravels of Rock Creek.

Granier's property had passed through several owners and one bankruptcy when E.T. Fisher leased the ground in 1932 and launched Atlantic City into one more minor gold boom. The gravel was carefully tested that summer and the willows burned off in the fall. The next spring a Cat with a dragline (like a crane with a scoop shovel) crawled a hundred miles northeast from the railroad at Rock Springs, a three-day trip at the blazing speed of one mile per hour. A movable processing plant was built on site, and by the summer of 1933 the dragline began eating its way upstream, first stripping the topsoil then biting mouthfuls of loose rock to drop into the processing plant which sorted the gold and shuffled the gravel out into neat, fifteen-foot-high conical piles. Rock Creek now looks like many of the old placer streams across the West, turned inside out into a corrugation of rock piles so even they might be beautiful if they weren't so unnatural.

Dredging was the next technical advance in placer mining after hydraulicking fell out of favor. It was ideal for the flat streams where hydraulics didn't work well. Invented in New Zealand, dredges were first used in America at Bannack, Montana, in 1897 but spread quickly to California, the Klondike, Boise Basin, Alder Gulch, the upper South Platte, and many other streams that had been either unworkable or "worked out" by older methods.

The traditional dredge was a floating barge with a bucket line (imagine an escalator with big scoops instead of stairs) that continually scraped up gravel and dropped it into a processor, excavating its own pond as it clawed its way upstream. Dredging would launch what became the twentieth-century trend in gold mining—the use of massive machinery to recover minute amounts of gold from tremendous volumes of earth.

The gravel-pile droppings of dredges are conspicuous and unmistakable years after the fact, but I was shocked one day to stumble upon the actual dredge that ate Rock Creek. I had snow-machined into my cabin the beginning of March, and four weeks later when I wanted to come out the machine was marooned on its own personal snow bank while the road had mostly melted down to dirt. So I slung on my backpack and marched across the sagebrush to Rock Creek, bucking three-foot drifts and a forty-mile-an-hour headwind.

It was nearing evening, with a quarter moon rising over the dark, twisted cliffs, when I walked into the dredge, a two-story wooden box on railroad wheels. There were gaping holes in the wood and tin siding, and a rickety stair that led up to the second floor. I climbed the stairs gingerly, swinging up on to beams to avoid the severed landing. Peering through the dim light of a leaky roof, I could see the sole occupant—an iron trommel four feet in diameter. The giant tube was punched with a gradation of holes, smaller at the top and larger at the bottom, which sorted gravel into five separate troughs to be washed according to size. A huge flywheel once turned the trommel, whose excrement fed out onto a belted boom that had made one last perfect, conical pile of cobbles. A hole in the roof marked where the hopper once fed the trommel, and an iron-lined sluice box, joining the collecting troughs, slumped down the back fifteen feet to the ground.

I crawled back down the stairs to find the only occupant below, the engine that once turned the fly wheel. Some of the timbers that framed the forty-by-twenty-foot structure were twelve inches square, all carefully hewn and pieced to design. Though the skin was peeling off, this skeleton would not melt back to brush anytime soon. I spent so much time crawling over the beast on that day in March that my walk to Atlantic City was completed in the dark by sketchy orb of headlamp.

The Fisher dredge washed more than eleven thousand ounces of gold from Rock Creek between 1933 and 1940, bringing one more boom to a district that always seemed a day late and a dollar short. It was a boom replicated in dozens of old mining districts across the West in the decade surrounding the Great Depression. High unemployment (which reached twenty-five percent in the trough of the depression) combined with a seventy-five percent increase in the price of gold by January of 1934 created both the push and the pull to launch a new gold rush.

Rusting bucket line dredges that had ground to a halt under the weight of inflation in the 1920s bobbed suddenly back to life when President Roosevelt demonetized gold and allowed the official price to float upward from $20 to $35 an ounce. In California, construction contractors put out of business by the depression moved their dragline excavators to the gold fields and began scooping gravel bars into washing plants like the one on Rock Creek. The number of placer mines nearly doubled to 1,784 in California between 1933 and 1934, most of them small operations with an output of less than fifty ounces. "Since the depression many people lacking other employment have panned for gold by hand in the streams of California with the hope of extracting enough to furnish the bare necessities of life," the Bureau of Mines reported in 1936, estimating that more than nineteen thousand were thus employed. "Although the average income of this class of miners has been pitifully small, a few of the more fortunate ones have picked up rich pockets."

On the Klondike, dozens of miners made a similar living by panning the dirt under Dawson's wooden sidewalks. In Montana, annual gold production increased 500 percent between 1932 and 1937, and the rivalry between Virginia City and Helena flared up again as a dredge in Alder Gulch vied with a dredge in Last Chance Gulch for largest gold producer in the state in 1936. As usual, Helena won.

At Deadwood in the Black Hills of South Dakota, it wasn't so much an increase in production as an increase in value that mattered. Anchored by the mighty Homestake Mine, annual gold production increased less than twenty percent between 1932 and 1935, but the value of that production went from around ten million to nearly twenty million dollars with gold's revaluation. Ore that had been too low in grade to mine was suddenly valuable, and old tailings were reworked for new wealth.

Colorado's boom began in downtown Denver, where hundreds flocked to work the mud flats and sand bars of the South Platte and Cherry Creek, making a dollar or less a day. Municipal officials encouraged this form of self-employment by offering streamside panning and sluicing schools. Hard rock mines reopened in the mountains, doubling Colorado's gold production. Bucket line dredges scraped the stream bed in downtown Breckenridge, and draglines mucked gravel in California Gulch, as well as Clear Creek, Boulder Creek, and the upper drainages of the South Platte including rivals Tarryall and Fairplay.

Nationwide, annual gold production had nearly tripled by 1940 to six million ounces. But with gold, a boom is inevitably followed by a bust. World War II brought a different national priority, and gold mining was made illegal to focus production on more essential materials. After the war, inflation reduced the monetary value of gold to an uneconomical level, and production dropped to a pre-Depression trickle. The Bretton Woods agreement in 1944 attempted to reestablish a standard where the American dollar was fixed to gold and other currencies were fixed to the dollar. The civilized world was unable to believe, as the economist Keynes believed, that gold was the barbarous relic of a medieval mindset.

The price of everything else began rising, while gold was still valued at $35 an ounce. Black market sales pushed the unofficial price upwards while governments dumped some of their reserves to try to force the price back down to the $35 level. Finally a two-tiered system was created where the market price was allowed to float while governments paid their debts at the official, depressed price. But in 1971, with a constant trade deficit draining U.S. reserves, President Nixon suspended international gold payments. Two years later the U.S. Senate approved free ownership of gold for the first time in forty years. The price began to rise like a hot air balloon, tripling by 1973 then doubling again by 1978. In 1980 the price rocketed up to $850 an ounce in a shark-like, speculative feeding frenzy, then crashed to less than $300 in 1982 and has been bobbing between $250 and $450 since.

Because gold is now a commodity, bought and sold like wheat or pork bellies, its value has become questionable. I was sitting in the saloon in Atlantic City on a day when the price of gold was at a twenty-year low, and heard the bartender say that gold as an investment was dead. "Buy now and shut up," I told him, and he should have taken my advice. While gold as an investment has a short history to make predictions from, the factors that affect the price have become very predictable. Low inflation and good returns in the stock market lead to low gold prices. High inflation and unstable currencies lead to rising gold prices, and the higher the price goes, the more people want to own it.

The problem with gold being sold as a commodity is that it doesn't behave like a commodity. It isn't consumed like wheat or burned like oil, it simply changes hands. Ninety percent of all the gold ever mined through history is still with us (some four billion ounces). The other ten percent continues to turn up—in galleon wrecks on the sea floor, in graves unearthed and stashes forgotten. And despite increased production worldwide, there has been no real glut on the market. People keep snatching it up as bracelets and necklaces in Mideastern *souks*, as bullion coins from government mints, in bricks and bars and paper contracts.

There has never been enough gold to go around, which is why it was finally demonetized. If the total world gold supply was divided up equally into the total world currency supply, an ounce of gold would have a book value of $200,000. Even Denver stockbrokers would be down on the banks of the Platte panning and sluicing their wages then.

But gold is no longer money, the bartender would argue. So why is it still valuable? At the very least, humans value it for their own adornment. Four million pounds a year of gold goes into jewelry, and it goes a long way. With 14 karat gold only half gold and half alloys, and most jewelry plated or "filled," people are often buying only the illusion of gold. But it is enough to make even a pauper feel like a princess.

Of the twenty-five hundred tons of gold mined in a year, four hundred tons get melted into bars, or stamped into coins and medallions to sell to investors. Three hundred tons are used by industry in a variety of applications. When electrical contacts are critical or prone to corrosion—on sensors that trigger airbags, for example, or computer chips and telephone jacks—gold is used to insure against failure. A gold coating on the windows of airplanes reflects heat both in and out, and can carry an electric current to defog and defrost. Gold-plated ovens help dry the water-based paints now used on automobiles, and gold-plated reflectors have been installed on Air Force One to confuse heat-seeking missiles.

But the real value of gold has always been measured in the long run. It is a "value of last resort." When economies crumble and currencies crash, gold is the surest hedge against an uncertain future. And while gold has been completely demonetized, governments still hang on to a third of all the gold ever mined, around forty thousand tons of it, their own hedge against an uncertain future. There has been a recent trend toward divestiture, with governments selling off some of their gold reserves, but it has been small as well as tentative. What governments get rid of, private citizens accumulate, and the balance between the two could be revolutionary. If gold is, indeed, a barbarous relic, we remain barbarians.

Chapter 13

Among twenty snowy mountains
The only moving thing
Was the eye of the blackbird.
from Wallace Stevens
"Thirteen Ways of Looking at a Blackbird"

STEVE GYORVARY owns a gold mine. You might think he is one of the luckiest people in the world. A standard reply to children begging for money used to be "What? Do you think I own a gold mine?" as if that circumstance would change everything. But Steve will be the first to tell you that owning a gold mine is not all it's cracked up to be.

His mine, the Mary Ellen, is perched on a ridge overlooking Rock Creek and the straggling head of Atlantic City. (It is one of those western ironies that most towns with the word "City" still attached to their names now have more phantom residents than real.) Atlantic City has always been a rebel town, refusing to die while its neighbors, South Pass City and Miner's Delight, Camp Stambaugh and Lewiston, passed quietly into ghostdom.

There are two saloons left in Atlantic City, and one church—a small log chapel with the door painted defiantly red. The saloons attract more converts. It was there I first met Steve Gyorvary, who is more likely to tell you and everyone else his opinions of the world at twelve midnight than at twelve noon when he appears shy, reflective, and too intelligent for his own sanity. Steve is thin and loose-jointed as a scarecrow, intractably unkempt, a generational fan of the Grateful Dead and gonzo journalist Hunter S. Thompson.

Steve invited me to take a tour of his gold mine, and I accepted. The Mary Ellen has the only permitted operating gold mill in the state of

Wyoming, as well as a long history of starts and stalls. Discovered in 1869, it joined the boom of the "Sweetwater Mines" as well as the bust that followed in 1875. Like a dozen other mines that line the ridges above Rock Creek—the Duncan up the ridge and Tabor Grand below, the Diana and Rose and Garfield across the creek—the Mary Ellen changed ownership every decade or two, selling for thousands of dollars one decade and hundreds the next as the degree of optimism dictated.

In the early 1890s the Mary Ellen was owned in partnership and worked by one of Atlantic City's more rebellious citizens: Rufus M. "Fatty" Ricketts. Born in Tennessee, Rufus Ricketts arrived in Atlantic City as a freighter in 1869 at age nineteen, and there he stayed through good times and bad. Ricketts was blue eyed, bland faced, and as his nickname suggests, increasingly girthed. A reporter for the local *Clipper* in 1877 called him "the jolly, wholesouled R.M. Ricketts, who is at present a gentleman of leisure, and owner of valuable mining property." But either the bad times began to wear on him, or he had a mean streak beneath his jolly exterior, or it was simply the rebellious town he lived in, but Fatty Ricketts had a penchant for trouble.

Civil court records show he was involved in a dozen lawsuits over twenty years. He sued the local storekeeper over a $3.50 cow hide in 1885. In 1893 his partner's wife had to sue Ricketts and her husband to get her "goods and chattels" (a side board, table, six chairs and rocking chair) so she could get away from them. Ricketts was sued three times for back wages and back pay for work done by various people, and had restraining orders issued against him by several mining partners.

Ricketts was arrested once for assault with intent to kill, and when the hydraulic mastermind Emile Granier was threatened in 1889 by a certain citizen in the streets of Atlantic City, and then begged the governor for troops to protect his life, whose name did I discover on a deed transferring a single cabin to Granier for the extortionary price of $450 at the same time Granier realized the troops weren't coming? It was indeed Rufus M. Ricketts.

But innocent-faced Ricketts wasn't a thug; he was a schemer. His outrages were carefully planned for his own gain. The assault charge against him, his only criminal court record, was no barroom brawl. He had his eye on the Rose Mine, owned in 1896 by Colorado silver king Horace Tabor. The Rose was an unpatented claim, which meant that every year a hundred dollars worth of assessment work had to be done to hold the claim or it would be open to relocation. Ricketts watched all year, staking a new location notice on the Rose to be ready to assume ownership. Then, on the last day of the year Ben Evans, hired by Tabor to do the annual assessment work, finally showed up at the Rose mine. Rufus Ricketts was waiting for him.

The headline of the *Clipper* for January 8, 1897, read: "BEN'S HEAD. Rufus M. Ricketts Works on it with a Winchester." Ben was knocked senseless, and when he came to he could see Rufus hurrying away to officially record the claim for himself. Ricketts pleaded not guilty to the charge of assault, and everyone but Ben must have thought it a good joke, for when the district court finally convened to try the case the following summer, Ricketts was fined a mere twenty-five dollars and court costs for assault and battery. There was some justice however, for Ricketts's scheme failed to gain him ownership of the Rose. It was still the property of H.A.W. Tabor the following year, and when Ben Evans went to do the annual assessment work in 1897, he took along a body guard just in case.

The Mary Ellen mine passed out of Ricketts's ownership in the mid 1890s, after he had removed several thousand dollars in gold ore and collapsed the main shaft in the process. Fatty's final record, his estate settlement, shows that as a "gentleman of leisure" Ricketts did pretty well for himself. When he died in 1904 at the premature age of fifty-four, he still owned a quarter interest in the Diana and Victoria Regina mines worth $1,500, and he left behind nearly $7,000 in cash. He probably wished he could take it with him.

The new owner of the Mary Ellen, a Denver capitalist named John Ibson, turned the mine over to the South Pass Mine and Milling Company for shares of stock, which the company then devalued trying to freeze Ibson out. The litigation that followed effectively closed the mine for more than a decade at the turn of the century. The South Pass Mine and Milling Company emerged from the lawsuit as owners in 1912, but within two years they had sold the Mary Ellen to the Beck Company that was working the Duncan mine next door. The Mary Ellen stayed attached to the Duncan for twenty years, shutting down in World War I, and reopening in the 1930s when gold mining was again profitable. But when the manager of the Duncan absconded with all the profits (including the miners' yearly wages) around 1937, the Mary Ellen was abandoned.

Two locals, Roy Cowden and Charlie Hall, relocated the Mary Ellen. Cowden sold out his half to Hall, and two years after Hall died in 1976, Steve Gyorvary's family bought the mine from Hall's widow. In its sporadic history, the Mary Ellen turned out sixty-six hundred ounces of gold. The ore was some of the richest in the district, but it was found in lenses that blossomed and pinched out unpredictably. When the Gyorvarys bought the mine, the 250-foot shaft was caving in on itself, and threatened to take the shaft house with it. Steve cleared out the four hundred tons of caved debris by hand, shoveling the muck into the bucket of a dragline, then crawling out of the shaft and into the dragline to empty each bucket onto the dump pile.

Steve had worked in the underground uranium mines of New Mexico until that industry crashed like a derailed locomotive in 1982. A geology degree from the University of Missouri had taught him about rocks, but he needed a jack-of-all-trades degree when he took over the Mary Ellen. After cleaning and retimbering the shaft, he pumped out the mine (flooding has been a constant and decisive problem at nearly all the South Pass mines) and mapped the remaining ore bodies. In 1985 Steve decided it was worth rebuilding the mill to process the remaining ore. It was a decision he came to regret.

It took five years and $60,000 to meet all the environmental regulations, post reclamation bonds and secure the permits to operate his mine and mill. The amount of land to be disturbed was less than a supermarket's parking lot. His return so far has been $0.00. When the Gyorvarys bought the Mary Ellen in the 1970s, there were quite a few small mine operators chipping away at dark, elusive gold veins. It wasn't impossible then for one man to own a gold mine. Now it is a corporate world: big machines, big ore bodies, big business. There is one advantage to being small. Large mines like the Homestake in South Dakota spend three hundred dollars or more to produce an ounce of gold. When the price of gold drops below $300, they lose money. Steve can turn out an ounce for a hundred dollars, and so can make a profit when his oversized counterparts cannot.

That is if he could get his mill running again. The day I went to visit he was in the middle of restructuring. The 1911 Denver Quartz Mill installed by the South Pass Mine and Milling Company, which greatly resembles an old Spanish arrastra—a wheel rolling around in a trough to crush ore—had been cast by the wayside in favor of a ball mill, which rolls the ore in a drum with various sized metal balls. The ball mill can crush three tons of ore per hour, while the old mill could only crush one ton. The "new" ball mill came out of the Duncan next door, and is well used. The jaw crusher, which gets the ore first, dates back to the mine's discovery. "We're riding into the twenty-first century on nineteenth century machines," Steve says.

Two-thirds of the equipment in a gold mill is made to pulverize ore, the other third to separate gold from grit. A mill on a hill is a good thing, so that gravity can move material for you. At the top of the Mary Ellen hill sits the shaft house, one of those tall and angular structures that make old mines so picturesque. The Mary Ellen has an inclined shaft following the vein down at an angle. You don't get a sense of vertigo like you do looking down a straight shaft, but it still appears ominously dark and slimy. Steve says that's what miners are all about—narrow veins, wet shafts, and ... well, I won't say what else. Ore from the shaft is wheeled to the mill building and dumped over a grizzly (a series of bars like a cattle guard to catch oversized chunks) into the jaw crusher. The jaw crusher reduces 70 percent of the ore into half-inch-minus pieces, which then go into a roll crusher (like a wringer) and come out three-eighths of an inch. The ball mill that grinds the ore further was still outside the mill building in a state of disassembly, as were several jigs to separate free-milling gold before it all got pulverized into powder and fed through a flotation system. The end product is a black powder that looks nothing like gold, but contains seventeen ounces of the precious metal in each ton. It gets shipped to a smelter for final recovery.

Further up the hill is a shack housing two diesel generators (salvaged from a construction company) that power all the machinery. To the side is the original shaft, caved in by Rufus Ricketts when he mined out the supporting pillars for their gold. The mine dump, underfoot everywhere, is a beautiful mosaic of colored stone—pink and beige, rust and tan. The original mill was so inefficient it didn't pay to work ore less than one ounce in value to the ton. They dumped the rest here in a gravelly rainbow fan, and it turns out I am walking on three-quarters of a million dollars worth of gold. At least as much gold ore is still below ground in the Mary Ellen, and it makes me think it might actually be nice to own a gold mine.

But all that gold won't buy Steve a haircut. He wouldn't be caught dead in a suit and tie. We all might get a round of beer at the Merc's saloon, but the way Steve lives has nothing to do with money. What he wants is to get up when he feels like it in the morning, drink his coffee in the old log cabin moved in from Camp Stambaugh that is now his mine office, and ponder what can be accomplished on this day. He cannot stand hypocrisy, and that includes becoming something he doesn't want to be. What he wants is to make a one-man gold mine work when no one else, including Atlantic City's saloon full of rebels, thinks he can. He might be a walking anachronism, but I like the shape of his footprints.

OCTOBER 16. Darkness comes with a finality these days, marking the end of summer. I still wish for daylight, for the ease and green of summer in a land somewhere on the edge of my mind, like a place I visited once and have almost forgotten. There is no softness here, no escape from the hard edges. Except, perhaps, at that certain time of day when dusk falls across the rolling hills, and sage takes the shape of folded velvet in a gray and silver hue.

I cannot tell you in an easy way why I like living here, in the middle of ten thousand empty acres. Except to say they are not empty. My neighbors are better than most, my back yard more interesting. At times I play Thoreau, working out the economics of my life—how many hours I spend cutting and splitting wood versus the number of hours I would have to work for someone else if I paid for central heating. It turns out that I'm not really "roughing it." My life is easier than most: no stress, no deadlines, no responsibility to anyone but myself.

And I learn here every day a lesson that is submerged under the artificial construction of our cities and suburbs, that is attempted but never realized in the textbooks of ninth-grade biology, and is obscured by the medieval mindset that the world revolves around us. I have found that the world is richer than we think, full of things that can delight and surprise us, things not of our own making, all connected by threads we have not woven. I would guess that is why people seek out wild lands, visit parks, watch birds, stop their cars to gawk at wildlife. It is all a glimpse into a larger world, one in which we are not at the center.

I have discovered that history is like that too, a web of stories in a circle of time. We can stand at one edge and look across to the distant other, and see fables; or put up a screen that filters truth, and view legends; or put on self-righteous glasses and see only immorality. But step into the circle and you see lives like your own, full of struggle and triumph, failure and hope, humanity and inhumanity. How much larger is your own life when you have lived so many others.

I believe that the past can bring scale to the present. We look back at where we've been, look ahead to where we might go, and come in a circle back to where we are. The wider that circle is, the larger our own humanity, and the smaller our inhumanity. History is not so much facts and figures (though these are endlessly useful for ascertaining truth and making comparisons), as it is stories with morals that teach us how to live. Without these stories we have no sense of consequence, and no perspective in our judgments.

It is all, in the end, a matter of scale and perspective. I was thinking the other day about how the Mary Ellen mine compared to the Homestake mine in the Black Hills where I once made a pilgrimage. It is like the difference between a homestead and a city, both in scale and perspective. Where the Mary Ellen employs one and the assorted specialist in welding or earth moving for a day or two, the Homestake employed 800 before low gold prices in the late 1990s forced that company to scale back to avoid losing too much money. The shaft of the Mary Ellen goes down 250 feet; the Homestake plunges 8,000 feet towards the center of the earth. The temperature just below the surface of the Homestake is 44 degrees; at 8,000 feet the temperature is 133 degrees. For every ton of ore mined at the Homestake, eleven tons of cooled, fresh air have to be circulated through two hundred miles of underground chambers on various levels.

Two hundred forty million kilowatt hours of electricity are consumed by the mine in a year, enough to power a number of South Dakota cities. One 400-ounce bar of gold requires the mining, hoisting, crushing, grinding, and cyanidation of more than two million pounds of ore. Around four hundred of those gold bars a year are poured at the Homestake's refinery, a military-looking building surrounded by concertina wire, with bullet-proof windows and armed guards. No one is told when or how the bars are shipped, or to whom, and a special alloy is added to the 99.8 fine gold so it can be identified if stolen.

The mine shafts at the Homestake are only five by eight feet square, which means the diesel-powered front-end loaders and all heavy machinery has to be lowered in pieces and reassembled underground. The shaft houses are proportionally tall, and the cables running the cages (for people and equipment) and skips (for ore) up and down the shaft are wound on gigantic cones to keep the cable from crossing itself. The hoist operator sits in a captain's chair and pushes a lever, far above the dust and darkness below. Two women were manning the hoists the day I was there, attentive but disconnected from the usual image of a miner. Hoisting used to be signaled by bells—one bell up if standing, stop if moving; two bells down; nine bells fire or cave in. The modern Homestake uses radio waves transmitted on the hoisting cable itself, and the only audible sound is the hum of large electric motors.

The Homestake is the oldest continuously operated gold mine in the world. Brothers Fred and Moses Manuel, with partners Alex Engh and Henry Hardy, discovered the vein in 1876. The Manuels hoped it would give them a stake to go home to Minnesota, which they had left nine years earlier to make their fortune in the West. They had tried Montana, but couldn't find a vein big enough to pay the transportation and development costs. They found more Indians than gold in Idaho. Moses then fought Apaches in Arizona, thinking a golden star might shine on some mountain in the southern desert. But the Apaches drove him out, and from Arizona Moses moved on to British Columbia, finding gold near Great Slave Lake. It was too cold for him in Canada, and he was back in Portland waiting passage to Africa when Custer's report of gold in the Black Hills showed up in the newspaper. Moses picked Fred up back in Helena, and headed for the Hills. They were among those who slipped in early, unconcerned about the Sioux, indifferent to the army blockade.

The Manuels located three good prospects in the spring of 1876, but hard rock claims were unlike placer bonanzas in that their value was neither obvious nor instantaneous. They built an arrastra to grind their ore, then a ten-stamp mill, and were finally making a little money when an agent for a California capitalist showed up and offered them $70,000 for the Homestake. The capitalist was George Hearst, who had a knack for buying the right mining property at the right time: Comstock silver, Anaconda copper, Black Hills gold (though in an uncharacteristic slip, he passed on Tombstone silver). The Manuels sold out for that $70,000 and returned home to Minnesota. The Homestake would produce $780 million in 1876 dollars, or more than 39 million ounces.

The Manuels stayed in Minnesota only long enough to find wives. Like thousands of others bitten by the gold bug, they went west again to seek yet another fortune. Moses died in a mine accident near Helena in 1905, and the less adventurous Fred eventually retired to Los Angeles. The town that grew up next to their Homestake mine was called Lead (pronounced leed, a miner's name for a vein, while the town of Leadville, Colorado is pronounced ledville for the lead ore that generated its silver bonanza). The Manuels' original mine shaft is now a giant hole in the ground, what they call the "Open Cut," on the edge of Lead's Main Street. Across a mountain of gold from the town of Deadwood, Lead's Main Street has been moved three times to accommodate the growing hole (from which has been excavated 288 million cubic yards of rock).

But the Open Cut is only the visible part of the Homestake mine, which delves into a huge underground ore body that was once a number of different mines that the Homestake acquired: the Oro Hondo and DeSmet, Golden Terra and Old Abe, Caledonia and Highland among others, bringing Homestake's total ground in the Black Hills to nearly twelve thousand acres. The Homestake's expansion and longevity are owed largely to cyanide. Production in all the Black Hills mines was grinding to a halt in the 1890s because the low grade ores couldn't be milled profitably. But the discovery in the late 1880s of the ability of sodium cyanide, in dilute solution combined with air and water, to remove almost all of the gold from refractory ores, allowed mines like the Homestake to suddenly double production and profitability. We think of cyanide as a deadly poison, but it is only poisonous when combined with a strong acid, like that found in people's stomachs. When combined with gold, it was the magical elixir that launched the twentieth century's biggest gold rush.

Most people aren't even aware of it—the gold rush was over long ago, they think. But the decade of the 1990s brought another gold boom to the American West. Seemingly overnight, Nevada is now the world's second largest gold producer, trailing only South Africa. Over eight million ounces of gold annually flow out of Nevada, more gold in one year than the ancient world could accumulate in fifteen centuries. It is "invisible" gold, missed by the old prospectors because the grains were so small they couldn't be seen in a pan, and would wash out anyway. Missed because it turned up in such unlikely rock—sedimentary limestone and siltstone where the gold was disseminated in a hot solution soaked up like a sponge. Missed in the dirt eroding from the sedimentary rock that can often be simply piled up and leached in a heap by the elixir of cyanide.

From deep in the ground at Homestake, mining has moved into infinitesimals in the Carlin Trend of northeastern Nevada, a geologic anomaly of microscopic gold particles infused into a broad band of 350-million-year-old sedimentary rocks during one of Nevada's many tectonic shake ups. If mountains are a good place to look for gold, Nevada should top the list, with more than 160 mountains ranges. The search to discover all of Nevada's treasure has taken a long time, and isn't over yet. The discoveries have come in waves, beginning with the splash of the Comstock that sent ripples out to wash over nearby Aurora in 1860, Unionville in 1861, and Austin in 1862. Eureka anchored a second silver boom in the 1870s, and was joined by Pioche, Tuscarora, and dozens of other smaller camps.

The mining boom was going bust by the turn of the nineteenth century, when wandering burros led Jim Butler to the rich silver veins of Tonopah. But gold was the real star of that new century. In 1904 thousands rushed to newly discovered Goldfield where mines were turning out $10,000 a day. In 1905 the stampede went south toward Death Valley to the gold of Rhylolite and Bullfrog, raising "a string of dust a hundred miles long." You might think that by the twentieth century the old prospectors would have abandoned their burros and their golden dreams to join the modern world, but Nevada has always lived in its own time zone.

So it is not surprising that this newest gold rush sprang up in Nevada on the eve of the twenty-first century. The Carlin Trend was actually discovered in 1961, with the first mine opened in 1965 by Carlin Gold Mining Company, which became Newmont Gold Company in 1986. Like any good gold rush, one discovery led to more discoveries, and two decades of development are finally paying off. Newmont is now the second largest gold producing company in the world, with nine open pits, four underground mines, and eighteen processing facilities in Nevada which turned out 2.5 million ounces of gold in 1999.

Newmont also owns gold mines in Mexico, Peru, Uzbekistan, and Indonesia. Gold mining has not only gone corporate, it has gone global. Foreign companies mine gold in America, and American companies like to hedge their bets overseas. Gold is as much a commodity as grain, and those who mine it, scraping earth from giant pits as level-bottomed as a wheat field, might be harvesting crops in oversized tractors.

Like their counterparts in the nineteenth century, many of these new gold rushers have come from earlier booms gone bust—uranium miners from Wyoming and New Mexico, copper miners from Utah and Arizona. One town is emptied and another filled to overflowing. Elko, Nevada, is the boom town at the center of the Carlin gold rush, its population tripled since 1980. Elko has made the best of its boom times, and is a fairly typical civic-minded small city of twenty thousand, set in the broad valley of the Humboldt River with surrounding hills of sage flanked by the ten-thousand-foot-high Ruby Mountains. Elko has managed to accommodate its swelling population because gold, as always, brings wealth.

The average miner's wage is $54,000 while other workers in the state average $30,000. But the second consistency in gold rushes is that boom is always followed by bust. The nearest shopping mall is a three-hour drive from Elko; no one wants to build a mall in a town that might someday have no shoppers. There are more than 470 ghost towns in Nevada, over 300 of which were once mining boom towns.

Mine geologists estimate ten to fifteen years of gold reserves still in the ground at the Carlin Trend, and they haven't quit looking for more. It is enough for the new gold rushers to pack up and move to Elko, buy a house, put their kids in school, live out some semblance of the American dream. And while some things haven't changed in this new gold rush—the accumulation of wealth, and the certainty that one day the source of that wealth will come to an end—other things have changed.

Many of the risks once inherent in the gold rush game have been minimized or eliminated. Corporations now core drill and measure and map ore bodies long before they begin mining. (In 1997 alone, forty-seven companies spent $131 million on exploration activities in Nevada.) And the physical risks gold rushers of the nineteenth century faced—the grueling overland crossings, the toil of pick and shovel, exposure to the elements and disease, the strain of uncertainty whether their labor would pay off, the separation from family, and the gambling of their lives—all are gone.

What is also gone is that chance in a lifetime, if you made it to the gold fields alive and were first in line to be in the right place at the right time, that one opportunity after buying a ticket in the gold rush lottery to see your name spelled out in glittering nuggets at the bottom of the pan, that possibility that you might, against all odds, succeed in the end and strike it rich.

At the beginning of the Black Hills gold rush, before the Homestake and other valuable discoveries at Deadwood, and before the Hills were even legally open to prospectors, Lieutenant Colonel Richard Irving Dodge was escorting a second geological survey to confirm Custer's report of gold. Dodge shook his head at the hordes of illicit miners pouring into the Hills and rooting for gold that was then only a rumor. "There is a fascination in the search after gold not to be accounted for," he observed. "For a dollar's worth of shining precious grains, a man will devote patience and labor which would have brought him three times the amount of money in almost any other business."

Dodge sagaciously predicted that "of each twenty men who will rush to the Black Hills as miners, nineteen would have been better off if they had remained at home. I am aware," he concluded, "that this statement will deter no man from going, as the American people are so constituted that each man expects himself to be the twentieth."

OCTOBER 23. I've been crouched by the water for an hour, spinning sand like an incantation. Four empty sample bags are crumpled on the stream bank, and I stand and stretch my tired muscles before I empty the fifth into my pan. I shrug at the sun-bleached wisps of hair tickling my cheek, and push back my battered Stetson to watch the dying afternoon light gild the canyon's cliffs. I bend again to wash the last panful of secretive earth, swirling sand and water in a sparkling gyration of suspense. From the last revealing spoonful I dab out tiny flakes of gold and drop them into the sample bottle which slides back into my pocket.

Reluctant to leave, I sit on the bank and dangle my booted feet in the clear Sweetwater. A plume of silt kicks up, then disappears downstream. It's growing colder in the shadows, and a milkdrop moon rises against the pale sky, three days from full. The hunter's moon. The wind gusting from the west rattles the dry willow branches. It is an autumnal wind, hollow with death. I listen to its variations, and wonder why it sings in a different key than the summer wind, why it lacks the crescendo of the winter gale.

And the light is an autumnal light. Oblique, half-hearted. It is again that time of year when the colors are surreal. I look, and look again, wondering if I have ever seen these shades before: bone and sinew. The landscape is an old carcass newly laid out for burial, and I am staying here to sing the dirges.

Notes on Sources

To be able to generalize, one must know an infinite number of particulars. The search for this minutiae has taken me to wider fields than might seem prudent, and I catalogued over two hundred sources in my pursuit to put events in a larger context. Many of the sources I read twice, and some three times to master both the story and the meaning of that story.

For a book that is mostly about western American history, I diverged from my subject frequently. But I could not comprehend the insanity of the California gold rush without placing it within humankind's long-standing fascination with gold, and the quest to gain it at any cost. Likewise, Custer's demise cannot be attributed simply to a bad military decision (which it was), but stretches back to a repeated cycle of advance and retreat on native lands beginning with the fur trade, and accelerated by the rush for gold.

There were several general sources that I used repeatedly for facts and details of the various rushes, and I will list them here to avoid repeating them later. Rodman Paul's *Mining Frontiers of the Far West, 1848–1880* was the best. T.H. Watkins's *Gold and Silver in the West: The Illustrated History of an American Dream*; and William Greever's *The Bonanza West: The Story of Western Mining Rushes, 1848–1900* were also useful.

Full bibliographic details for books will appear in the Bibliography, but I will list the details for periodicals and manuscripts in this Notes on Sources, and will leave them out of the Bibliography to relieve the amount of repetition. Sources I consulted but did not quote may appear only in the Bibliography.

I frequently translated the value of gold between the nineteenth and twentieth centuries to give some perspective on what it was really worth. Because the price of gold now fluctuates daily, I based my translation on the average between which gold has fluctuated in the last decade, settling on a value of $350 for an ounce of gold. In the nineteenth century, an ounce of raw gold was worth around $18. It would be interesting to compare the purchasing power between now and then, but I will forebear from that one last diversion.

Chapter 1

Sources for the Oregon Trail are divided between three chapters. Here I used Unruh, Goetzman, and Franzwa. Horace Greeley's quote comes from his account of *An Overland Journey from New York to San Francisco in the Summer of 1859.*

Ghost Dog Ridge is unnamed on the maps, and is so called by me because one night while camped at its base I heard a dog barking, and there was no one nearby. Maybe it was a coyote, or some sheepherder around the ridge. … The geology and gold content of that area comes from Love, J.D., Antweiller, J.C., and Mosier, E.L. "A New Look at the Origin and Volume of the Dickie Springs–Oregon Gulch Placer Gold at the South End of the Wind River Mountains." *Thirtieth Annual Field Conference.* Wyoming Geologic Association Guidebook, 1978.

Chapter 2

Robert Frost's poem about fences is called "Mending Wall," from *North of Boston.* The history of Lewiston and the discovery of South Pass gold comes from Coutant; Bancroft; Pfaff; and Murray, Robert A. "Miner's Delight, Investor's Despair." *Annals of Wyoming* 44 (Spring 1972); and Nickerson, H.G. "Early History of Fremont County." *Quarterly Bulletin,* Wyoming State Historical Department 2 (July 1924).

Rossiter W. Raymond's "Statistics of Mines and Mining in the States and Territories West of the Rocky Mountains." *Annals of Wyoming* 40 (October 1968) repeats the 1869 *Sweetwater Mines* story of the 1842 Georgian fur trapper; and Todd Guenther includes the 1860 Denver news in "Why Were They Here? Or Gold, the Cariso, & South Pass City." *Wind River Mountaineer* 7 (April–June 1991). *The Lewiston Gold Miner,* May, 1894 was the only issue printed of a newspaper designed to promote the mines there.

The story of the Mormon emigration to Utah is told in Hafen, *Handcarts to Zion*; and Stegner, *The Gathering of Zion.* John Chislett's account is from "Mr. Chislett's Narrative" in Stenhouse. Kate B. Carter collected details of the handcart pioneers; and Rebecca Cornwall described the rescue of the handcart companies.

Chapter 3

Most of the Indian prehistory comes from Frison's *Prehistoric Hunters of the High Plains.* Virginia Trenholm gives the best history of the local Shoshones; and details of the native acquisition of horses comes from Roe.

Hafen's *Mountain Men and Fur Traders of the Far West* is the most complete biography of those hearty souls. For information on the fur trade I also consulted Frost; Sandoz; Billington; and Morgan.

It is interesting to compare Washington Irving's somewhat fictionalized *Astoria* with the journal accounts of Robert Stuart in both Spaulding and

Rollins (a book that actually has more footnotes than text!). The Ruxton quote comes from *Life in the Far West*; and Decker's from *The Diaries of Peter Decker.*

Chapter 4

Sir Leonard Woolley's *Excavations at Ur* is a fascinating story of early archeology, as is Howard Carter's *The Tomb of Tutankhamen.* I spent months trying to track down the oldest golden ornament ever found, but failed in the end. If anyone has that information, please let me know. ... Other golden trivia comes from Allen; and Willsberger. The early colonial attempts to find gold are from Smith's *The Generall Historie of Virginia.*

John Sutter's account of his momentous meeting with James Marshall in January of 1848 is told in Lewis, *Sutter's Fort: Gateway to the Gold Fields,* as are the details of Sutter's life. Descriptions of Marshall come from Caughey; Martinez; and William Johnson's *The Forty-Niners.* Henry Bigler gives an account from Marshall of the meeting, as well as a description of what went on at Sutter's Mill in Erwin Gudde's *Bigler's Chronicle of the West.*

Walter Colton's account of events is presented in Gilmore. Details of the snowballing effects of the discovery come from Caughey; and Rodman Paul's *California Gold: The Beginning of Mining in the Far West.* James Carson told his tale in *Recollections of the California Mines*; and Edward Buffum in *Six Months in the Gold Mines.*

Most of the details of Ted Hurst and Sarah Gillespie are local gossip that I cannot vouch for, but there is an interesting file in the Geologic Survey of Wyoming (MR 44-5 "Big Nugget Placer Claims") marked "Confidential" that comments on the fractiousness of Mrs. Gillespie, and she wrote columns for the Lander newspaper detailing events at Radium Springs and nearby Lewiston.

Chapter 5

The history of alchemy comes from Cummings; and Redgrove; and the contributions of alchemy to modern science and details of the transmutation of bismuth to gold are from Brock's *The Norton History of Chemistry.*

Eastern America's first response to the news of gold in California comes from Caughey, and Donald Dale Jackson's *Gold Dust*, a highly entertaining account of a nation's undoing. John Banks's story is presented in Scamehorn's *The Buckeye Rovers in the Gold Rush.*

General information on the California Trail was drawn from Mattes and Unruh. Statistics on death and illness are in Rieck, Richard L. "A Geography of Death on the Oregon–California Trail, 1840–1860." *Overland Journal* 9 (1991); and Olch, Peter D. "Treading the Elephant's Tail: Medical Problems on the Overland Trails." *Overland Journal* 6 (1988).

The precocious travel writer Bayard Taylor rushed back to the States and published *Eldorado*. A description of mining techniques can be found in Young. The decline of California mining is described in Rodman Paul's *The Far West and the Great Plains in Transition, 1859–1900*. The stories of Mary Jane Megquier, Luzena Wilson, and the anonymous pie baker come from JoAnn Levy's *They Saw the Elephant: Women in the California Gold Rush*, an account that gives a different perspective of the experience.

There is some mystery behind who Louise Clappe really was. She signed her letters "Dame Shirley," and though Clappe might also be a pseudonym, I have retained it because it is easiest to find her *The Shirley Letters from the California Mines, 1851-1852* under that authorship.

Chapter 6

History of the Burr mine comes from Pfaff; and Rosenberger, Robert G. "The Lewiston Mining District." *Archeological and Historical Survey of Mine Reclamation Sites in Fremont County, Wyoming*. South Pass Historical Files. The geology of South Pass was gleaned from Bailey; Hausel; and Lageson. The geology of gold came from Park; Hutchison; and Riley, but do not blame them for my interpretation. The story of the lucky Colorado miners in the Depression comes from Voynick. Phoebe Gustin's obituary from the *Wyoming State Journal* (10/15/64) contained the story of Lewiston's blizzards.

Mark Twain's *Roughing It* is one of his most entertaining books, describing his stagecoach ride west, and his experiences in Virginia City. The darker side of that experience is revealed in Clemens, Samuel Langhorne, *Mark Twain's Letters*. Other details of Clemens's life are from Benson; and Kaplan. Twain's fellow journalist, Dan DeQuille, also gives an entertaining account of events in *Big Bonanza*. J. Ross Brown tells his version in *A Peep at Washoe and Washoe Revisited*.

Chapter 7

The history of gold as money was drawn from Sherman; Berman; Moore; and Ritter. The myth of gold at the end of the rainbow is from Gnadinger. The economics of money was interpreted from Kemmerer; and Friedman.

George Jackson's diary, edited by Hafen, appears in *Colorado Magazine*, 12 (November 1935). Early gold discoveries in Colorado, including Luke Tierney's account of the Cherokee/Russell party of 1858, are told in Hafen's *Pike's Peak Gold Rush Guidebooks of 1859*. The response to those discoveries is collected in Hafen's *Colorado Gold Rush: Contemporary Letters and Reports 1858–1859*. The stories of those who took to the trails, including J. Heywood; the Blue brothers; E.H.N. Patterson; Charles Post; and A.D. Richardson are found in Hafen's *Overland Routes to the Gold Fields, 1859*.

Other details of the Colorado rush come from Voynick; Brown; Fossett; and Muriel Wolle's *Stampede to Timberline*. Reverend John Dyer told his own story in *The Snow-shoe Itinerant*; and that of Father Machebeuf is told by W.J. Howlett. Ovando Hollister revisited the old camps in *The Mines of Colorado*; as did that now-seasoned reporter Bayard Taylor in *Colorado: A Summer Trip*.

Colorado's Indian war is documented in Nadeau; Grinnell; and Duane Smith's *The Birth of Colorado*. The attack of Black Kettle's camp by Chivington's troops is told in Stan Hoig's *The Sand Creek Massacre*.

Chapter 8

Hervey Johnson's letters home are collected in Unrau's *Tending the Talking Wire: A Buck Soldier's View of Indian Country 1863–1866*. Caspar Collins's biographer is Agnes Wright Spring. Other details of the 1860s Indian war are from Nadeau; Hyde; Trenholm; and McDermott. The fight at Platte Bridge Station where Caspar Collins lost his life is well described in J.W. Vaughn's *The Battle of Platte Bridge*. Sgt. Isaac Pennock's account appears in Hebard.

The "Journal of Henry Edgar—1863" appears in *Contributions to the Historical Society of Montana* 3 (1900); as does Peter Ronan's "Discovery of Alder Gulch." James Stuart gives his version of events in "The Yellowstone Expedition of 1863." *Contributions to the Historical Society of Montana*. 1 (1876); which also includes Sam Hauser's comments. In that same issue is Granville Stuart's "A Memoir of the Life of James Stuart." A larger account of their life in Montana is Granville Stuart's *Forty Years on the Frontier*. Additional biographical details come from Treece.

Robert Raymer's *Montana: The Land and the People* gives background to the Montana gold rushes, as does Dorothy Johnson's excellent book, *The Bloody Bozeman*. Elizabeth Fisk's letters are collected in the Montana Historical Society Archives (Manuscript Collection 31. Fisk Family Papers, 1858–1901). Additional details of her life are from Petrik. Helena's larger history is from Palmer; and Schreiner, M. Murray. "Last Chance Gulch becomes the Mountain City of Helena." *Montana Magazine of History* 2 (October 1952).

Professor Thomas Dimsdale's account of *The Vigilantes of Montana* may be biased in favor of the "good guys." R.E. Mather tries to balance that in *Hanging the Sheriff: A Biography of Henry Plummer.* The haunted story of galloping hoofbeats is repeated from Dorothy Johnson, which also includes one version of the miners' joke about leaving Heaven in a gold rush to Hell.

Chapter 9

See Chapter 2 Notes for sources on the discovery of South Pass. Harriet Loughary's journal appears in Holmes (Volume 8). Details of life and business in South Pass City are drawn mostly from its two newspapers: the *Sweetwater Mines* published in 1868; and the *South Pass News* printed from 1869 to 1870. Additional details are from Marion Huseas's *Sweetwater Gold*; and James Sherlock's *South Pass and its Tales.*

James Chisholm's account, *South Pass, 1868: James Chisholm's Journal of the Wyoming Gold Rush,* is worth reading simply for the pleasure of its literacy. The details of Tom Quinn's end are from that volume's editor, Lola Homsher.

The continuation of the Indian War, with Red Cloud closing the Bozeman Trail, is described in Nadeau; Olson; and Hebard. Also consulted are Smith, Sherry L. "The Bozeman: Trail to Death and Glory." *Annals of Wyoming* 55 (Spring 1983); and McDermott, John D. "Price of Arrogance: The Short and Controversial Life of William Judd Fetterman." *Annals of Wyoming* 63 (Spring 1991). Margaret Carrington tells her harrowing tale in *Ab-sa-ro-ka: Land of Massacre.*

The repercussions of Red Cloud's victory, washing over South Pass, are evident in contemporary newspaper accounts, and are told by Nickerson (op. cit. Notes Chapter 2) and John R. Murphy in "The Indian Raids of 1869 and 1870." *Wind River Mountaineer* 9 (July–September, 1993). Lt. Stambaugh's demise and biography are given in Delo, David. "Camp Stambaugh ... the miner's delight." *Wind River Mountaineer* 3 (October–December, 1987).

There has been an interesting historical controversy over what role Esther Morris played in winning suffrage for Wyoming women. That controversy is discussed in Scharff, Virginia. "The Case for Domestic Feminism: Woman Suffrage in Wyoming." *Annals of Wyoming* 56 (Fall 1984); and Massie, Michael A. "Reform is Where You Find It: The Roots of Woman Suffrage in Wyoming." *Annals of Wyoming* 62 (Spring 1990).

Chapter 10

For background sources on the Oregon–California Trail, see Chapter 1 and Chapter 5 Notes. Sir Richard Burton's exposé of the American West is *City of the Saints, and Across the Rocky Mountains to California.* Fremont's description of Sweetwater Canyon is from Jackson, *The Expeditions of John Charles Fremont.* Biographical details come from Egan's *Fremont: Explorer for a Restless Nation.*

I perused more than forty emigrant journals to see what they had to say about crossing South Pass. The emigrant who commented on "the most crooked, hilly, stony road" up Rocky Ridge was Susan Cranston, in Holmes (Volume 3). Lodisa Frizzell's heartbreaking journal is *Across the Plains to California in 1852.* Amelia Hadley, in Holmes (Volume 3), "found plenty of snow in some ravines." William Johnston described bad weather in *Experiences of a Forty-Niner.* Lucena Parsons, in Holmes (Volume 2) thought the road was "as broad & fine as any turnpike;" and Margaret Frink, in Holmes (Volume 2), found "it was not easy to tell when we had reached the exact line of the divide." Details of climate and the Little Ice Age are from Pielou.

The 1874 expedition to the Black Hills led by Custer is chronicled in Donald Dean Jackson's *Custer's Gold.* Newspaper correspondence from that expedition is collected in Krause, *Prelude to Glory.* Annie Tallent described her illicit quest for gold in *The Black Hills, or The Last Hunting Ground of the Dakotahs.* Details of the Black Hills rush are from Watson Parker's *Gold in the Black Hills.*

The mystery of Ezra Kind's 1834 stone epitaph is presented in Thomson's *The Thoen Stone: A Saga of the Black Hills.* Richard Hughes describes his adventures in *Pioneer Years in the Black Hills.* Hickok's life is detailed in Rosa, *They Called Him Wild Bill.* Estelline Bennett tells her own entertaining story in *Old Deadwood Days.*

The Battle of the Little Big Horn, featuring "Custer's Last Stand," is probably one of the most famous military episodes in American history. For various reasons, confusion reigns as to what exactly happened there. I have drawn upon more recent studies that shed an interesting light on events, namely Gray, whose *Custer's Last Campaign* includes a time analysis to place Custer and his troops; and Richard Fox's *Archaeology, History, and Custer's Last Battle,* which brings in new evidence from archeological surveys completed after wild fires cleared the battle ground. For the Indian point of view, see Kammen, Marshall, and Lefthand.

Chapter 11

Ed Schieffelin's story is presented by Underhill in *The Silver Tombstone of Edward Schieffelin.* Details of the rush to Tombstone are from Myers; and Faulk. Paula Mitchell Marks gives a detailed analysis of events surrounding the "Shootout at the OK Corral" in *And Die in the West.*

The story of Cripple Creek gold is most entertainingly told in Marshall Sprague's *Money Mountain.* Another book I highly recommend is Pierre Berton's *The Klondike Fever,* from which I drew largely for the stories of Carmack and Henderson, Barry and Ladue, as well as Belinda Mulroney.

Will Shape's adventures are told in *Faith of Fools: A Journal of the Klondike Gold Rush.* Polly Purdy's name when she wrote her memoirs in *My Ninety Years,* was Martha Louise Black. Melanie Mayer describes routes to the Klondike, as well as the women newspaper correspondents, and gives slightly different details on Belinda Mulroney. E. Hazard Wells contributes his version of events in *Magnificence and Misery.*

Chapter 12

Rivera's rediscovered journal appears in Jacobs, G. Clell. "The Phantom Pathfinder: Juan Maria Antonio de Rivera and His Expedition." *Utah Historical Quarterly* 60 (Summer 1992). Escalante's journal has been edited by Bolton in *Pageant in the Wilderness: The Story of the Escalante Expedition to the Interior Basin, 1776.* The Sarraceno quote comes from Sanchez, *Explorers, Traders, and Slavers: Forging the Old Spanish Trail 1678–1850.* The legend of the "Lost Rhoads Mine" is repeated by George Thompson, without documentation, in *Faded Footprints: The Lost Rhoads Mines and other Hidden Treasures of Utah's Killer Mountains.*

The story of the *Wasatch Wave* reporter in 1897 is quoted from "Those Old Spanish Gold Mines." *Utah Historical Quarterly* 9 (July, October 1941). William Palmer's comments are in "Other Spanish Mines." *Utah Historical Quarterly* 9 (July, October 1941). Details of Utah mineral discoveries are from Stokes, William Lee. "Geology of Utah." *Utah Museum of Natural History Occasional Paper Number 6,* 1987. The *Salt Lake Herald* report of 1871 is quoted from Thompson's *Some Dreams Die: Utah's Ghost Towns and Lost Treasures.* Also consulted was Hill, Joseph J. "Spanish and Mexican Exploration and Trade Northwest from New Mexico into the Great Basin, 1765–1853." *Utah Historical Quarterly* 3 (January 1930).

Spanish exploration in the Americas is drawn from Horgan; and Innes. Details of Cortés and Pizarro are from Prescott's *History of the Conquest of Mexico and History of the Conquest of Peru.* Coronado's story is told by Bolton in *Coronado: Knight of Pueblos and Plains.*

Emile Granier details are from Pfaff; the *Fremont Clipper*; and Noble, Bruce J., Jr. "A Frenchman in Wyoming: The South Pass Mining Misadventures of Emile Granier." *Wyoming Annals* 65 (Winter 1993–94). The history of hydraulic mining comes from Rohe, Randall E. "Hydraulicking in the American West." *Montana: The Magazine of Western History* 3 (Spring 1985). The legal battle over California hydraulic mining is told in Robert Kelley's *Gold vs. Grain.*

Details of the Fisher Dredge on Rock Creek are pieced together from Pfaff; and articles in the *Wyoming State Journal,* including Peter Sherlock's "Active Mining Operations at Atlantic City and South Pass" (4/11/1935). Background information on the history of dredging is from Murray, Sharon. "A Bucket Full: Boise Basin's Dredging Heritage, 1898–1951." *Idaho Yesterdays* 34 (Fall 1990).

Statistics of gold production in the Depression are found in Bureau of Mines. *Minerals Yearbook.* Washington: United States Government Printing Office, 1933–1940. Details of Depression mining in Colorado are from Voynick; and in the Klondike from Berton. The economic and investment potential of gold is drawn from Friedman; Sherman; Green; and Gotthelf. Industrial use of gold was supplied by The Gold Institute, "Facts About Gold." http://www.goldinstitute.org (8/19/99).

Chapter 13

The rebellious Rufus Ricketts managed to stay largely out of history despite his antics. I pieced his story together from U.S. Census records, and claim and district court records at the Fremont County Courthouse. There is a brief file on his family at the Lander museum (thanks to Todd Guenther for his help on this and other matters!), and a picture of Ricketts is in Coutant's *History of Wyoming.*

The history of the Mary Ellen mine is drawn from Pfaff; Beeler, Henry C. "A Brief Review of the South Pass Gold District, Fremont County, Wyoming." *Wyoming State Geologist Report* 1908; Jamison, C.E. "Geology and Mineral Resources of a Portion of Fremont County, Wyoming." *Wyoming State Geologist Bulletin No. 20,* 1926; and from personal communication with Steve Gyorvary.

The discovery of the Homestake mine is told in Parker Watson's *Gold in the Black Hills,* and details of that mine's development, including the influence of cyanide, is in the same author's *Deadwood: The Golden Years.* Facts and figures about the Homestake are from Lin Carr's *History of the Homestake Gold Mine*; and the Homestake Mining Company. "Operations Fact Files, 1999." http://www.homestake.com (9/3/99).

The Carlin Trend is described in Stewart, John H. "Geology of Nevada." *Nevada Bureau of Mines and Geology Special Publication 4,* 1980; and Lapointe, Daphne D., Tingley, Joseph V., and Jones, Richard B. "Mineral Resources of

Elko County, Nevada." *Nevada Bureau of Mines and Geology Bulletin 106*, 1991. Production figures are from Nevada Mining Association. "Mining & Economics." http://www.nevadamining.org (4/24/00).

Details of the Newmont corporation can be found at Newmont Mining. "Investor Relations." http://www.newmont.com (4/24/00); and Mining Technology. "Carlin Trend Newmont Gold Company." http://www.mining-technology.com (11/15/99). Nevada's past rushes are described in Elliott, Russell R. *History of Nevada*; and Paher, Stanley W. *Nevada Ghost Towns & Mining Camps*. For a look at the effects of Nevada's latest gold rush, see Grabowski, Richard B., Raney, Russell G., and Wetzel, Nicholas. "Behind Nevada's Golden Renaissance." *Minerals Today* (December 1991); and Cauchon, Dennis. "Nevada's New Gold Rush." *USA Today*. (7/21/97).

The quote from Richard Dodge comes from his *The Black Hills*; and the story behind the story of his Black Hills expedition is told in Kime, Wayne R., ed. *The Black Hills Journals of Colonel Richard Irving Dodge*.

I want to add one more note on how I put this book together, for when I started I had no plan of what it would contain (could you tell?). I had only a subject, gold, and a curiosity of where it would lead me. What ended up in the book was what I found interesting. It's a long ride from beginning to end, but I hope the view has been worth the trip.

Bibliography

Allen, Gina. *Gold!* New York: Thomas Y. Crowell Co., 1964.

Bailey, Richard W.; Proctor, Paul Dean; and Condie, Kent C. "Geology of the South Pass Area, Fremont County, Wyoming." *Geological Survey Professional Paper 793.* U.S.G.S., 1973.

Bancroft, Hubert Howe. *History of the Pacific States of North America.* Vol. 20. San Francisco: The History Company, 1890.

Bennett, Estelline. *Old Deadwood Days.* Lincoln: University of Nebraska Press, 1982.

Benson, Ivan. *Mark Twain's Western Years.* Stanford: Stanford University Press, 1938.

Berman, Allen G.; and Malloy, Alex G. *Warman's Coins and Currency.* Radnor: Wallace-Homestead Book Co., 1995.

Berton, Pierre. *The Klondike Fever.* New York: Alfred A Knopf, 1958.

Billington, Ray Allen. *The Far Western Frontier.* New York: Harper & Brothers, 1956.

Black, Martha Louise. *My Ninety Years.* Anchorage: Alaska Northwest Publishing Co., 1976.

Bolton, Herbert E. *Coronado: Knight of Pueblos and Plains.* Albuquerque: University of New Mexico Press, 1949.

———, ed. *Pageant in the Wilderness: The Story of the Escalante Expedition to the Interior Basin, 1776.* Salt Lake City: Utah State Historical Society, 1972.

Brock, William H. *The Norton History of Chemistry.* New York: W.W. Norton & Co., 1992.

Brown, Robert L. *Ghost Towns of the Colorado Rockies.* Caldwell: Caxton Printers, 1968.

Browne, J. Ross. *A Peep at Washoe and Washoe Revisited.* Balboa Island: Paisano Press, 1959.

Buffum, E. Gould. *Six Months in the Gold Mines.* 1850. Reprint. The Ward Ritchie Press, 1959.

Burton, Richard Francis, Sir. *City of the Saints, and Across the Rocky Mountains to California.* New York: Harper and Brothers, 1862.

Carr, Lin. *History of the Homestake Gold Mine.* Lead: Homestake Mining Co., 1994.

Carrington, Margaret Irwin. *Ab-sa-ro-ka: Land of Massacre.* Philadelphia: J.B. Lippincott & Co., 1879.
Carson, James H. *Recollections of the California Mines.* 1852. Reprint. Oakland: Biobooks, 1950.
Carson, Phil. *Across the Northern Frontier: Spanish Explorations in Colorado.* Boulder: Johnson Books, 1998.
Carter, Howard. *The Tomb of Tutankhamen.* New York: E.P. Dutton, 1972.
Carter, Kate B. "Records of the Handcart Pioneers." *Our Pioneer Heritage.* Vol. 14. Salt Lake City: Daughters of Utah Pioneers, 1971.
———. "The Handcart Pioneers." *Treasures of Pioneer History.* Vol. 5. Salt Lake City: Daughters of Utah Pioneers, 1956.
Caughey, John Walton. *Gold is the Cornerstone.* Berkeley: University of California Press, 1948.
Chisholm, James. *South Pass, 1868: James Chisholm's Journal of the Wyoming Gold Rush.* Lincoln: University of Nebraska Press, 1960.
Clappe, Louise Amelia Knapp Smith. *The Shirley Letters from the California Mines, 1851–1852.* New York: Alfred A. Knopf, 1949.
Clemens, Samuel Langhorne. *Mark Twain's Letters.* Vol. 1. New York: Harper & Brothers Publishers, 1917.
Conference on the History of Western America. *Probing the American West.* Santa Fe: Museum of New Mexico, 1962.
Cornwall, Rebecca, and Arrington, Leonard J. *Rescue of the 1856 Handcart Companies.* Provo: Charles Redd Center for Western Studies, 1982.
Coutant, C.G. *History of Wyoming.* Laramie: Chaplin, Spafford & Mathison, 1899.
Cummings, Richard. *The Alchemists.* New York: David McKay Co., 1966.
Decker, Peter. *The Diaries of Peter Decker.* Georgetown: Talisman Press, 1966.
DeQuille, Dan [pseud.]. *The Big Bonanza.* Apollo edition. New York: Alfred A. Knopf, 1969.
Dimsdale, Thomas J. *The Vigilantes of Montana.* Norman: University of Oklahoma Press, 1953.
Dodge, Richard Irving. *The Black Hills.* New York: James Miller, Publisher, 1876.
Doty, Richard G. *Money of the World.* New York: Grosset & Dunlap, 1978.
Dyer, John L. *The Snow-shoe Itinerant.* Cincinnati: Cranston & Stowe, 1890.
Egan, Ferol. *Fremont: Explorer for a Restless Nation.* Garden City: Doubleday & Co., 1977.
Elliott, Russell R. *History of Nevada.* Lincoln: University of Nebraska Press, 1987.
Faulk, Odie B. *Tombstone: Myth and Reality.* New York: Oxford University Press, 1972.

Fossett, Frank. *Colorado: Its Gold and Silver Mines, Farms and Stock Ranges, and Health and Pleasure Resorts.* New York: C.G. Crawford Printer, 1879.
Fox, Richard Allan, Jr. *Archaeology, History, and Custer's Last Battle.* Norman: University of Oklahoma Press, 1993.
Franzwa, Gregory M. *The Oregon Trail Revisited.* St. Louis: Patrice Press, 1972.
Friedman, Milton; and Schwartz, Anna Jacobson. *A Monetary History of the United States, 1867–1960.* Princeton: Princeton University Press, 1963.
Frison, George C. *Prehistoric Hunters of the High Plains.* 2d rev. Ed. New York: Academic Press, 1991.
Frizzell, Lodisa. *Across the Plains to California in 1852.* New York: New York Public Library, 1915.
Frost, Donald McKay. *Notes on General Ashley, The Overland Trail and South Pass.* Barre: Barre Gazette, 1960.
Frost, Lawrence A. *The Custer Album.* New York: Bonanza Books, 1984.
Frost, Robert. "Mending Wall." *North of Boston.* New York: Henry Holt and Company, 1915.
Gilmore, N. Ray, and Gilmore, Gladys, eds. *Readings in California History.* New York: Thomas Y. Crowell Co., 1966.
Gnadinger, Louise. "Myth, Magic and Alchemy" in *Gold.* New York: Alpine Fine Arts Collection, 1981.
Goetzmann, William H. *Exploration and Empire.* New York: Vintage Books, 1972.
Gotthelf, Philip. *The New Precious Metals Market.* New York: McGraw-Hill, 1998.
Gray, John S. *Custer's Last Campaign.* University of Nebraska Press, 1991
Greeley, Horace. *An Overland Journey from New York to San Francisco in the Summer of 1859.* New York: Knopf, 1964.
Green, Timothy. *The Prospect for Gold: The View to the Year 2000.* New York: Walker and Co., 1987.
Greever, William S. *The Bonanza West: The Story of Western Mining Rushes, 1848–1900.* Norman: University of Oklahoma Press, 1963.
Grinnell, George Bird. *The Fighting Cheyennes.* Norman: University of Oklahoma Press, 1956.
Gudde, Erwin G. *Bigler's Chronicle of the West.* Berkeley: University of California Press, 1962.
Hafen, LeRoy R. *Mountain Men and Fur Traders of the Far West.* 10 vols. Glendale: Arthur H. Clark Co., 1965–1972.
———., ed. *Colorado Gold Rush: Contemporary Letters and Reports 1858–1859.* Southwest Historical Series. Vol. 10. Glendale: The Arthur H. Clark Co., 1941.

———., ed. *Overland Routes to the Gold Fields, 1859*. Southwest Historical Series. Vol. 11. Glendale: The Arthur H. Clark Co., 1942.
———., ed. *Pike's Peak Gold Rush Guidebooks of 1859*. Southwest Historical Series. Vol. 9. Glendale: The Arthur H. Clark Co., 1941.
———.; and Hafen, Ann W. *Handcarts to Zion*. Glendale: The Arthur H. Clark Co., 1960.
———.; and Rister, Carl Coke. *Western America*. New York: Prentice-Hall, 1941.
Hausel, W. Dan. "Gold Districts of Wyoming." *Report of Investigations No. 23*. Geological Survey of Wyoming, 1980.
Hawkes, Jacquetta. *History of Mankind*. Vol. 1. New York: Harper & Row, 1963.
Hayden, F.V. *Eleventh Annual Report of the United States Geological and Geographical Survey of the Territories embracing Idaho and Wyoming, 1877*. Washington: Government Printing Office, 1879.
Hebard, Grace Raymond; and Brininstool, E.A. *The Bozeman Trail*. Cleveland: Arthur H. Clark Co., 1922.
Hoig, Stan. *The Sand Creek Massacre*. Norman: University of Oklahoma Press, 1961.
Hollister, Ovando J. *The Mines of Colorado*. New York: Promontory Press, 1974.
Holmes, Kenneth L., ed. *Covered Wagon Women: Diaries and Letters from the Western Trails, 1840–1890*. 8 vols. Glendale: Arthur H. Clark Co., 1983.
Horgan, Paul. *Conquistadors in North American History*. New York: Farrar, Straus and Co., 1963.
Howlett, W.J. *Life of the Right Reverend Joseph P. Machebeuf*. Pueblo: The Franklin Press Co., 1908.
Hughes, Richard B. *Pioneer Years in the Black Hills*. Glendale: The Arthur H. Clark Co., 1957.
Huseas, Marion McMillan. *Sweetwater Gold*. Cheyenne: Cheyenne Corral of Westerners International Publishers, 1991.
Hutchison, Charles S. *Economic Deposits and their Tectonic Setting*. New York: John Wiley & Sons, 1983.
Hyde, George E. *Red Cloud's Folk*. Norman: University of Oklahoma Press, 1937.
Innes, Hammond. *The Conquistadors*. New York: Alfred A. Knopf, 1969.
Irving, Washington. *Astoria*. Clatsop edition. Portland: Binfords & Mort, nd.
Jackson, Donald Dale. *Gold Dust*. New York: Alfred A. Knopf, 1980.
Jackson, Donald Dean. *Custer's Gold: The United States Cavalry Expedition of 1874*. New Haven: Yale University Press, 1966.
Jackson, Donald, and Spence, Mary Lee, eds. *The Expeditions of John Charles Fremont*. Vol. 1. Urbana: University of Illinois Press, 1970.

Johnson, Dorothy M. *The Bloody Bozeman.* Missoula: Mountain Press Publishing Co., 1983.

Johnson, William Weber. *The Forty-Niners.* Alexandria: Time-Life Books, 1974.

Johnston, William G. *Experiences of a Forty-Niner.* New York: Arno Press, 1973.

Kammen, Robert; Marshall, Joe; and Lefthand, Frederick. *Soldiers Falling into Camp: The Battles at the Rosebud and the Little Big Horn.* Encampment: Affiliated Writers of America/Publishers, 1992.

Kaplan, Justin. *Mark Twain and His World.* New York: Simon and Schuster, 1974.

Kelley, Robert L. *Gold vs. Grain: The Hydraulic Mining Controversy in California's Sacramento Valley.* Glendale: The Arthur H. Clark Co., 1959.

Kemmerer, Donald L. "The Role of gold in the Past Century" in *Gold is Money,* edited by Hans F. Sennholz. Westport: Greenwood Press, 1975.

Kime, Wayne R., ed. *The Black Hills Journals of Colonel Richard Irving Dodge.* Norman: University of Oklahoma Press, 1996.

Krause, Herbert, and Olson, Gary D. *Prelude to Glory: A Newspaper Accounting of Custer's 1874 Expedition to the Black Hills.* Sioux Falls: Brevet Press, 1974.

Lageson, David R.; and Spearing, Darwin R. *Roadside Geology of Wyoming.* Missoula: Mountain Press Publishing Co., 1988.

Levy, JoAnn. *They Saw the Elephant: Women in the California Gold Rush.* Norman: University of Oklahoma Press, 1992.

Lewis, Oscar. *Sutter's Fort: Gateway to the Gold Fields.* Englewood Cliffs: Prentice-Hall, 1966.

Marks, Paula Mitchell. *And Die in the West.* New York: William Morrow and Co., 1989.

———. *Precious Dust: The Saga of the Western Gold Rushes.* New York: William Morrow & Co., 1994.

Martinez, Lionel. *Gold Rushes of North America.* Secaucus: Wellfleet Press, 1990.

Mather, R.E.; and Boswell, F.C. *Hanging the Sheriff: A Biography of Henry Plummer.* Salt Lake City: University of Utah Press, 1987.

Mattes, Merrill J. *The Great Platte River Road.* Lincoln: University of Nebraska Press, 1987.

Mayer, Melanie J. *Klondike Women: True Tales of the 1897–1898 Gold Rush.* Athens: Swallow Press, 1989.

McDermott, John D. *Dangerous Duty: A History of Frontier Forts in Fremont County, Wyoming.* Fremont County Historic Preservation Commission, 1993.

Moore, Carl H.; and Russell, Alvin E. *Money: Its Origin, Development and Modern Use.* Jefferson: McFarland & Co., 1987.

Morgan, Dale L., ed. *The West of William H. Ashley.* Denver: The Old West Publishing Co., 1964.

Myers, John Myers. *The Last Chance: Tombstone's Early Years.* New York: E.P. Dutton & Co., 1950.

Nadeau, Remi. *Fort Laramie and the Sioux Indians.* Englewood Cliffs: Prentice-Hall, 1967.

Olson, James C. *Red Cloud and the Sioux Problem.* Lincoln: University of Nebraska Press, 1965.

Paher, Stanley W. *Nevada Ghost Towns & Mining Camps.* Berkeley: Howell-North Books, 1970.

Palmer, Tom. *Helena: The Town and the People.* Helena: American Geographic Publishing, 1987.

Park, Charles F., Jr., and MacDiarmid, Roy A. *Ore Deposits.* 3d ed. San Francisco: W.H. Freeman and Co., 1975.

Parker, Watson. *Deadwood: The Golden Years.* Lincoln: University of Nebraska Press, 1981.

———. *Gold in the Black Hills.* Norman: University of Oklahoma Press, 1966.

Paul, Rodman. *California Gold: The Beginning of Mining in the Far West.* Cambridge: Harvard University Press, 1947.

———. *Mining Frontiers of the Far West, 1848–1880.* New York: Holt, Rinehart and Winston, 1963.

———. *The Far West and the Great Plains in Transition, 1859–1900.* New York: Harper & Row, 1988.

Petrik, Paula. *No Step Backward: Women and Family on the Rocky Mountain Mining Frontier, Helena, Montana, 1865–1900.* Helena: Montana Historical Society Press, 1987.

Pfaff, Betty Carpenter. *Atlantic City Nuggets.* Private Printing, 1978.

Pielou, E.C. *After the Ice Age: The Return to Life of Glaciated North America.* Chicago: University of Chicago Press, 1991.

Prescott, William H. *History of the Conquest of Mexico, and History of the Conquest of Peru.* New York: The Modern Library, 1931.

Raventon, Edward. *Island in the Plains: A Black Hills Natural History.* Boulder: Johnson Books, 1994.

Raymer, Robert George. *Montana: The Land and the People.* Chicago: The Lewis Publishing Company, 1930.

Redgrove, H. Stanley. *Alchemy: Ancient and Modern.* Chicago: Ares Publishers, 1980.

Riley, Charles M. *Our Mineral Resources.* New York: John Wiley & Sons, 1967.

Ritter, Lawrence S.; and Silber, William L. *Money.* New York: Basic Books, 1973.

Roe, Frank Gilbert. *The Indian and the Horse.* Norman: University of Oklahoma Press, 1955.

Rollins, Philip Ashton, ed. *The Discovery of the Oregon Trail: Robert Stuart's Narratives of his Overland Trip Eastward from Astoria in 1812–13.* Lincoln: University of Nebraska Press, 1995.

Rosa, Joseph G. *They Called Him Wild Bill.* Norman: University of Oklahoma Press, 1964.

Ruxton, George Frederick. *Life in the Far West.* Norman: University of Oklahoma Press, 1951.

Sanchez, Joseph P. *Explorers, Traders, and Slavers: Forging the Old Spanish Trail 1678-1850.* Salt Lake City: University of Utah Press, 1997.

Sanders, Helen Fitzgerald. *A History of Montana.* Chicago: The Lewis Publishing Company, 1913.

Sandoz, Mari. *The Beaver Men.* New York: Hastings House, 1964.

Scamehorn, H. Lee, ed. *The Buckeye Rovers in the Gold Rush.* Athens: Ohio University Press, 1989.

Schlissel, Lillian. *Women's Diaries of the Westward Journey.* New York: Schocken Books, 1982.

Scott, Douglas D., and Fox, Richard A., Jr. *Archaeological Insights into the Custer Battle.* Norman: University of Oklahoma Press, 1987.

Shape, William. *Faith of Fools: A Journal of the Klondike Gold Rush.* Pullman: Washington State University Press, 1998.

Sherlock, James L. *South Pass and its Tales.* New York: Vantage Press, 1978.

Sherman, Eugene J. *Gold Investment Theory and Application.* New York: Simon and Schuster, 1986.

Smith, Duane A. *The Birth of Colorado.* Norman: University of Oklahoma Press, 1989.

Smith, John. *The Generall Historie of Virginia.* 2 vols. Glasgow: J. Maclehose and Sons, 1907.

Spaulding, Kenneth A., ed. *On the OregonTrail: Robert Stuart's Journey of Discovery.* Norman: University of Oklahoma Press, 1953.

Sprague, Marshall. *Money Mountain: The Story of Cripple Creek Gold.* Boston: Little, Brown and Co., 1953.

Spring, Agnes Wright. *Caspar Collins.* New York: AMS Press, 1967.

Stegner, Wallace. *The Gathering of Zion.* New York: McGraw-Hill Book Co., 1964.

Stenhouse, T.B.H. *The Rocky Mountain Saints: A Full and Complete History of the Mormons.* New York: D. Appleton and Co., 1873.

Stuart, Granville. *Forty Years on the Frontier.* Glendale: The Arthur H. Clark Co., 1967.

Tallent, Annie D. *The Black Hills, or The Last Hunting Ground of the Dakotahs.* St. Louis: Nixon-Jones Printing Co., 1899.

Taylor, Bayard. *Colorado: A Summer Trip.* New York: G.P. Putnam and Son, 1867.

———. *Eldorado.* New York: Alfred A. Knopf, 1949.

Thompson, George A. *Faded Footprints: The Lost Rhoads Mines and other Hidden Treasures of Utah's Killer Mountains.* Salt Lake City: Roaming the West Publications, 1991.

———. *Some Dreams Die: Utah's Ghost Towns and Lost Treasures.* Salt Lake City: Dream Garden Press, 1982.

Thompson, Robert Luther. *Wiring a Continent.* Princeton: Princeton University Press, 1947.

Thomson, Frank S. *The Thoen Stone: A Saga of the Black Hills.* Detroit: The Harlo Press, 1966.

Treece, Paul Robert. *Mr. Montana: The Life of Granville Stuart, 1834–1918.* Ph.D. dissertation, Ohio State University, 1974.

Trenholm, Virginia Cole; and Carley, Maurine. *The Shoshonis: Sentinels of the Rockies.* Norman: University of Oklahoma Press, 1964.

Twain, Mark. *Roughing It.* Signet Classic Edition. The New American Library, 1962.

Underhill, Lonnie E. *The Silver Tombstone of Edward Schieffelin.* Tuscon: Roan Horse Press, 1979.

Unrau, William E., ed. *Tending the Talking Wire: A Buck Soldier's View of Indian Country 1863–1866.* Salt Lake City: University of Utah Press, 1979.

Unruh, John D., Jr. *The Plains Across.* Chicago: University of Illinois Press, 1979.

Vaughn, J.W. *The Battle of Platte Bridge.* Norman: University of Oklahoma Press, 1963.

Voynick, Stephen M. *Colorado Gold: From the Pike's Peak Rush to the Present.* Missoula: Mountain Press Publishing Co., 1992.

Watkins, T.H. *Gold and Silver in the West: The Illustrated History of an American Dream.* New York: Bonanza Books, 1971.

Webb, Walter Prescott. *The Great Frontier.* Austin: University of Texas Press, 1951.

Wells, E. Hazard. *Magnificence and Misery.* Garden City: Doubleday & Co., 1984.

Willsberger, Johann. *Gold.* Garden City: Doubleday, 1976.

Wolle, Muriel Sibell. *Stampede to Timberline.* Boulder: Muriel S. Wolle, 1949.

———. *The Bonanza Trail: Ghost Towns and Mining Camps of the West.* Bloomington: Indiana University Press, 1953.

Woolley, Sir Leonard. *Excavations at Ur.* New York: Thomas Y. Crowell Co., 1965.

Young, Otis E., Jr. *Western Mining.* Norman: University of Oklahoma Press, 1970.

Index